3D Concrete Printing Technology
Configuration with Green and Self-Healing Concrete

Dr. Tejwant Singh Brar

Dr. Mohammad Arif Kamal

Shubham Singh

Published by **Materials Research Forum LLC**
Millersville, PA 17551, USA

Published as part of the book series
Materials Research Foundations
Volume 134 (2022)
ISSN 2471-8890 (Print)
ISSN 2471-8904 (Online)

Print ISBN 978-1-64490-214-1
ePDF ISBN 978-1-64490-215-8

Distributed worldwide by

Materials Research Forum LLC
105 Springdale Lane
Millersville, PA 17551
USA
http://www.mrforum.com

Printed in the United States of America
10 9 8 7 6 5 4 3 2 1

Table of Contents

Preface

The construction techniques and materials have evolved from the early days of mud construction when concrete was not developed and was considered an unconventional material. On the other hand, the same concrete is today a conventional and common constructional material, signifying that the world is constantly evolving. As a result, it's no surprise that concrete has become one of the most widely used substances in modern civilization after water. It is evident that concrete is required in almost every type of construction, but are we utilizing its maximum potential? The future demands a more sustainable approach to building techniques and materials. When paired with emerging future technologies, concrete may exhibit to be much more efficient in terms of performance, life span, and carbon emissions. Among the future building, technologies are 3D printing, which the world is now experimenting with. While it will take time for this automated technology to substitute the conventional methods, it is the one that will eventually dominate the construction industry's momentum.

India is one of the leading users of concrete, it is the 2nd largest cement manufacturing nation in the world owing to this the carbon footprint is reaching heights, accounting for 38% emissions, which is creating global warming and climate change. The greenhouse gases are eroding the ozone layer and worsening the air quality. As a nation, we have an urgent need to take severe steps to cut emissions, which may be accomplished partly by using sustainable and eco-friendly practices, notably in the building sector. The study is based on a comparison between traditional construction techniques (namely cast-in-situ, pre-stress, and post-tension) and 3D printing concrete construction using four primary parameters: mechanism, composition, time, and cost. The study enables an understanding of the fundamental operation of each technology, as well as their advantages and disadvantages, which facilitates the formulation of comparison criteria for all the techniques. This book '3D Concrete Printing Technology: Configuration with Green and Self-Healing Concrete' contains Seven chapters that focuses on finding the potential perks and drawbacks of using 3D printing concrete technology in the current scenario. The study is an aggregated analysis that allows us to argue the need for an urgent transition to sustainable building techniques, particularly in India. 3D concrete printing is one of India's most promising future industries; the technology seems to have the potential to transform the whole construction industry. This book shall be beneficial to architects, civil engineers, building scientists, academicians and construction industry professionals.

CHAPTER 1

Introduction

1. Introduction

The construction industry has seen various transformations of techniques to materials, starting from the early era of mud construction when concrete was not invented and was considered to be an unconventional material. While on the other hand, the same concrete is now a conventional and common material, whereas now the world is aiming towards further evolution. Today concrete is the second most-consumed substance on earth after water by us humans, our man-made world is built on concrete. So, we see it being used in almost every form of construction, but if we question ourselves are we even utilizing the full potential of concrete? The future awaits a sustainable form of construction techniques and materials, concrete when combined with upcoming future technology may prove to be much more efficient in terms of performance, life span, less carbon release, etc. Among the future construction methods, one is the 3D printing technology, which the world right now is slowly experimenting upon. This truly automated technique will take time to replace the traditional ways, but slowly and steadily they are the ones that will overtake the momentum of the construction industry (Betsky, 2014). 3D printing also known as additive manufacturing is one of the most efficient forms of construction. It has better productivity, low cost, faster construction, produces less waste, has a lesser carbon footprint, and even gives geometric freedom in designing (Lageman, 2019).

Similarly, 4d printing is another kinetically evolved technique that is currently under research. The difference between 3D printing and 4d printing is that 3D printing refers to the layer-by-layer construction and henceforth produces a static structure. While 4d printing is a mixture of smart materials having rigid and expandable properties, they tend to change their shape under underexposure to light, temperature, and other natural factors (Unknown, 2020). So, if we are shifting towards a better way of constructing shouldn't the material aspect also evolve? Talking about the different sustainable materials such as 3D composites, super-thin steel, solar concrete, Hempcrete, etc are the ones that would eventually set new standards of sustainability. But their fusion with 3D printing needs to be looked at since the technology right now is evolving itself in context to the structural need of the world, and hence our focus bends towards more structurally stable material. Therefore, to support this style of construction a new form of concrete evolved further i.e., the Digital concrete. This form of concrete is in a fluid state which when discharged from the nozzle hardens.

It is highly used in 3D printing construction, where it does not require steel rebar reinforcement, rather millions of steel fiber particles are the ones that are mixed with it and hence provide resistance to tension (Wangler, et al., 2019). Other structural futuristic materials are self-healing and green concrete, where self-healing is the one when exposed to water produces calcite and self-heals the cracks in the structure, hence reducing infrastructure maintenance requirements and cutting greenhouse gases (Zitzman, 2018). While Green concrete is composed of mixing cement with one of the industrially produced waste, therefore helping in waste management and reducing carbon emission. So, the overall research area focuses on how the upcoming technology will eventually replace the traditional way of construction and proceed to a green and sustainable future along with a greater material strength hence getting mankind ready for future natural disasters.

A building structure is made up of many different components, including the foundation, walls, windows and doors, ceiling, roof, plumbing, and wiring, and no single business prints entire houses. When we say 'to print a building' we just mean printing the walls of a building because the printer hasn't yet participated in other operations. The 3D printing construction has the potential to largely replace traditional construction. Traditional stone construction with blocks and bricks is currently the most endangered. A printed house is no different from a concrete house because the materials used in printing houses comprise a blend of cement and sand (Chen et al, 2015). The printer is nothing more than a device for laying down the material. The sole difference is that the printer is a machine or robot, an automated system that does not require human intervention. Figure 1 shows the Museum of the future, the first 3D printed building of Dubai.

Figure 1. Museum of the future, the first 3D printed building of Dubai

2.   Justification and Need for the Study

Today India is considered a developing nation, according to the statistics India will become a superpower nation by 2035. This means India by then would have become efficient and powerful in various sectors such as business, automobile, stock market, food, and transportation and to be a superpower it should have a good infrastructure that is highly dependent on our construction industry. The industry must rapidly innovate, it should experiment to adopt various forms of sustainable materials and technology. Why is it so that the US is using certain construction technique which has not yet come to India? Why do we not see sustainable materials being used on-site such as green concrete? This may be due to cost issues or maybe due to lack of awareness, affordability, structural strength, etc.

India is the third-largest greenhouse gas emitter, under which the construction industry accounts for a major hold in releasing carbon dioxide in the production of concrete, cement, etc. So, going by the reports we can say that we must start to revolutionize our industry and innovate and promote the use of sustainable forms of materials and techniques. Talking about the material aspect, various sustainable materials are slowly replacing the use of conventional materials such as bricks. For example, super-thin steel, solar concrete façade cladding, bamboo, hempcrete bricks, and Timbercrete are materials that are on top in terms of sustainability. These materials are highly used in countries outside but not that much in India. So, instead of talking about various futuristic materials, the focus is to research the sustainable forms of structural materials within the Indian context (Betsky, 2014). Figure 2 shows the worldwide forecast of 3D printing technology in manufacturing and construction industry.

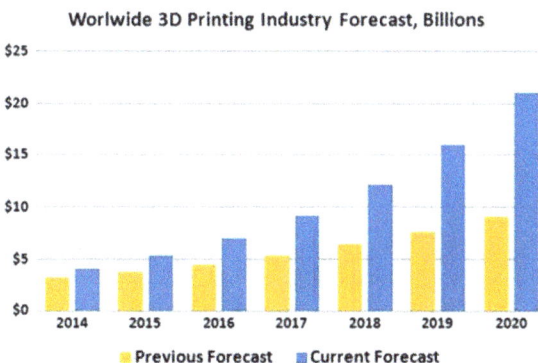

Figure 2. The worldwide 3D printing industry forecast

Today concrete is the second most-consumed substance on Earth, it is one of the strongest structural materials. India is the second-largest producer of cement in the world and hence the use of concrete is the primary material in any infrastructure development. The research therefore would focus on different sustainable forms of concrete evolution in India and how they can be made viable to be used on actual sites rather than only in terms of experimentation. 3D printing or concrete contouring is another rapidly growing construction technique in the world though it is yet to hit the construction industry of India, it is one of the most eco-friendly forms of construction i.e. reduces waste production, consumes less energy, consumes less material, etc. So, the research focuses to bridge the gap and establish a connection between the sustainable form of concretes using green and self-healing concrete with 3D printing technology. Right now 3D printing is limited to prototyping units and construction using conventional concrete, but the research here would try to fuse 3D printing with eco-friendly concrete along with challenges faced with its actual usage of it within the Indian Context (Lageman, 2019).

3. Objectives of the Study

To understand the broader impact of using sustainable forms of concrete alongside 3D printing technology based on the future of the Indian Construction Industry. The main objectives of this study are summarized below:

- To individually analyze the material integrity and challenges with such sustainable forms of concrete
- To establish the core need for using sustainable materials on construction sites
- To analyze the principle and challenges faced in 3D printing construction and the use of such forms of concrete
- To analyze the different mix designs we need to provide the material to sync it with 3D printing technology.
- To establish the benefits and structural behavior of the fused material with 3D printing

4. Scope and Limitations

3D printing is a technology that in recent times has been the mass production factor in the prototyping industry. But its constant growth has now started to reach the hands of the construction Industry. It not only provides eco-friendly construction but also reduces the cost and time of construction. The research contains an experimental and analytical

Materials Research Forum LLC
https://doi.org/10.21741/9781644902158

approach that can somehow or somewhere contribute to tackling the current striking issues of climatic and environmental hazards caused by construction.

The research somewhat has a limitation in terms of producing primary research only w.r.t. experimenting with 3D printing due to lack of technology in India. Also currently, there is a lack of experimentation knowledge that needs to be carried out to provide evidence for the research i.e., the need for a technical person who can educate about the various experiments that can be performed as primary evidence.

5.   Research Methodology

The research area is based upon a practical approach, so to prove the study the methodology would be based upon a comparative analysis using secondary research data. The paper would begin by establishing the evolution of construction techniques over the years and how it has impacted different sectors Then there would be a comparative analysis between 3D printing technology and the different forms of conventional construction techniques based upon certain criteria common in both. These criteria are cost, time of construction, mechanism, and composition. This study would help us analyze and lay down our argument about which technology is more efficient. Then this same criterion will be put forward for the analysis of green and self-healing concrete with conventional concrete. The study would give us two separate elements now, which will then be studied by fusing each other and the efficiency will be observed. The research would then focus on how this can revolutionize the Indian Construction market The research would then end upon concluding and gathering inferences from the analysis.

6.   The Expected Outcomes

The research is expected to raise awareness in the construction industry about the use of sustainable technology and material on a large scale. It tends to analyze the challenges faced by the market in adopting such methods and aims to provide certain key solutions to overcome the negligence of the usage of such methods on actual construction sites. This study is based on a comparison between traditional construction techniques (namely cast-in-situ, pre-stress, and post-tension) and 3D printing concrete construction using four primary parameters: mechanism, composition, time, and cost. The study enables an understanding of the fundamental operation of each technology, as well as the advantages and disadvantages of each, which facilitates the formulation of comparison criteria for all the techniques. This book focuses on finding the potential perks and drawbacks of using 3D printing concrete technology in the current scenario.

The study would discuss and bring upon the limitations of using sustainable materials at a large scale and a deep study about the 3D printing technology and the way it can uplift the Indian construction market. This book would eventually provide a study about the permuted mix designs of sustainable material that can be easily and economically fused with 3D printing technology. It tends to impart cognizance and an urge to adopt new construction methods and materials as an alarming gesture towards the changing climatic conditions and henceforth taking mankind to somewhat a sustainable future. The 3D concrete printing is one of India's most promising future industries; the technology seems to have the potential to transform the whole construction industry.

CHAPTER 2

Conventional Construction Technology

1.   Introduction

Construction technologies are methods that are used by human civilization to colonize, build and develop. We cannot deny the fact that technology is rooted in the past; it is dominating the present and extends to the future. Everything we use or consume is due to the technological advancements we have achieved. In the early times, the buildings were dominated by natural materials such as soil, wood, and stone, and today we see concrete as the widely consumed/produced material for construction. Why we have jumped to concrete? What are the other natural and man-made materials that can be used either in place of concrete or with sync in concrete? For this, we need to travel back in time and start from the beginning when natural materials were first used as constructing materials. The Industrial Revolution had a major impact that drastically altered the methods and technologies, promoting steel, glass and concrete (Wu, et al., 2019).

2.   Evolution of Construction Technology

The advancement in construction technology was not only just in terms of materials but also there was subsequent development in the mechanized equipment that is used today i.e., the widespread use of modern construction equipment had contributed to the development of environmental control technologies. Building structures constructed earlier were usually based upon the geological environment and approach, but today with the rapid development of technology and information, modern technology has almost ablated the geographical concept of time and space, resulting in the convergence of architecture (Wu, et al., 2019). The building innovation is joined by the rise of structures, and the total history of building innovation conveys the whole improvement of the structure. The building structure is the fundamental component of building innovation, and it very well may be utilized to support the historical backdrop of building technology. As indicated by records, building structure has existed since the start of human development and has been affected by numerous viewpoints like the economy, legislative issues, science and innovation, and culture. It is separated into the following stages:

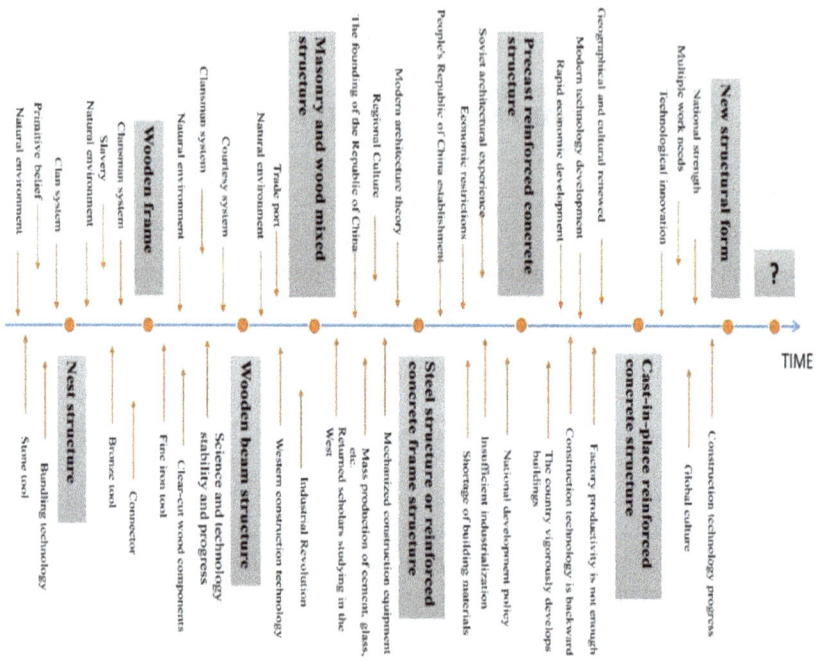

Figure 1. Time diagram of the development of building technology

The very first structure that emerged was the nest structure about 1.7 million years ago to around the 21st century BC. This era was the birth of natural construction materials such as soil and wood. Our ancestors have a rich experience of labor and therefore they eventually moved to ground buildings and dry column buildings (Wu, et al., 2019). In slave society (around 21st century BC to around 2nd century BC), the grouping of freedoms and the rise of bronze instruments had advanced the improvement of the wooden edge structure and the plate-building strategy (Figure 1).

3. Cast-in-Situ Construction Technology

Cast-in-place concrete sometimes referred to as poured-in-place concrete, is a method of concrete that is carried out in situ, or in the completed placement of the concrete component. For concrete slabs and foundations, as well as components like beams, columns, walls, and roofs, cast-in-place concrete is the best option. The concrete is usually delivered to the job site in an unhardened condition, usually by a ready-mix concrete truck. The concrete is placed in the desired spot or into a dumper or pump through a chute that extends from the rear of the vehicle. While cast-in-place concrete offers more flexibility and adaptability, it may be difficult to regulate the mix, especially if the weather isn't cooperating. Cast-in-place concrete needs a strength test and curing period, making it more time-consuming to produce than precast concrete. Since the structural system has fewer joints, less handling equipment is needed.

3.1   Composition/Mix Design in Cast-in-Situ Construction

The mix design for conventional concrete usually used for Cast-in-Situ is M20 Grade whose nominal mix ratio is 1:1.5:3 (Cement: Sand: Aggregate) with a compressive strength of $20 N/mm^2$. However, engineers may sometimes adjust the aforementioned ratios to meet their own needs. For instance, they may raise the sand content of concrete to improve its workability or lower the cement content to minimize the cost of concrete. However, excessive sand and cement content change will have a detrimental effect on the strength of concrete. As a result, it is generally not recommended to raise the fine aggregate content (river sand or M sand) by more than 30% over the permitted ratios. Cement and sand will initially fill the spaces between the coarse aggregate before assuming their area when mixing the materials. As a result, the total dry volume of cement, sand, and aggregate required for concrete preparation should be more than the entire volume of concrete (Table 1). As a result, 1.57 cubic meters of 'total dry volume' of cement, sand, and aggregate are required to make 1 cubic meter of M20 concrete. The total dry volume is according to the government rate (Standard, 2007, 2009).

3.1.1   Cement, Sand and Coarse Aggregate Requirement for M20 Grade Concrete

The nominal Concrete Mix Ratio of M20 grade concrete is 1:1.5:3.

Cement = 1 Part

Sand = 1.5 Part

Aggregate = 3 Part

Total dry volume of ingredients required = 1.57 cu. mtrs.

Table 1. Mix design of 1 cubic meter of M20 grade concrete

The volume of cement needed	= Ratio of cement x1.57/ (1+1.5+3)
	= 1x 1.57/5.5 = 0.285 cu. M
The volume of sand needed	= Ratio of Sand x 1.57 / (1+1.5+3)
	= 1.5x 1.57/5.5 = 0.427 cu. M
The volume of aggregate needed	= Ratio of aggregate x 1.57/(1+1.5+3)
	= 0.427 x 32 = 0.854 cu. M

3.1.2 Weight of Cement Required for 1 Cubic Meter of M20 Grade Concrete

The needed cement weight is derived by multiplying the cement volume by the cement bulk density. PPC and OPC cement have an average bulk density of 1440 kg/cu.m (Kumar, n.d.). Therefore the number of cement bags required is 8 bags (Table 2).

Table 2. Weight of cement needed for 1 cubic meter of M20 grade concrete

Weight of cement needed	= Volume of cement x bulk density of cement
	= 0.285 cu. m x 1440 kg/cu. m
	= 410 kg
No. of cement bag required	= 410 /50
	= 8.2 bags

3.1.3 Volume of Sand and Aggregate Required for 1 Cubic Meter of M20 Grade Concrete

The volume of sand and aggregate for 1 cubic meter of M20 grade concrete is shown in Table 3.

Table 3. Volume of sand and aggregate for 1 cubic meter of M20 grade concrete

The volume of sand needed	= Volume of sand x 35.32
	= 0.427 x 35.32 = 15.08 cu. ft.
The volume of aggregate needed	= Volume of aggregate x 35.32
	= 30.16 cu. ft.

3.1.4 Volume of Sand and Aggregate Required for 1 Cubic Meter of M20 Grade Concrete

Suppliers commonly sell sand and coarse aggregate in CFT and UNITS.

- 1 UNIT = 100 cubic feet
- 1 cum =35.32 cubic feet

A cubic meter of M20 concrete requires 15.08 cft sand and 30.16 cft aggregate (Table 4).

Table 4. Mix quantities of 1 cubic meter of M20 grade concrete with different units

Grade	Cement	Sand	Aggregate
M20	$0.25\ m^3$	$0.427\ m^3$	$0.854\ m^3$
M20	410 kg	15.01 cu. ft.	30.16 cu. ft.
M20	8.2 bags	15.01 cu. ft.	30.16 cu. ft.

Also, the initial setting time required for the Cast-in-Situ technique is 24-48 hours and after 28 days of curing it achieves its full strength this shows that time is a constraint here that delays the construction (Albrecht, 2021).

3.2 Cost Analysis of Cast-in-Situ Construction

The cost analysis includes material, labor, overhead, water, and contractor profit. The specifics of all operations involved in the activity should be provided for rate analysis. Concrete rate analysis summarizes material costs, labor overhead, contractor profit, and water rates. It includes the following:

(a) Quantity and cost of material (cement sand aggregate and water quantity and their cost)

(b) Water charges (1 tanker)

(c) Overhead expenses (3%)

(d) Cost of equipment and tools (2%)

(e) Contractor profit (10%)

(f) Labor rate and cost

3.3 Rate Analysis for 1 cum Concrete of M20 (1:1.5:3)

The cement, sand, and aggregate ratio in M 20 concrete is 1:1.5:3, with 1 part cement, 1.5 sand, and 3 aggregate (Civilsir, 2021). The processes for calculating the amount and cost of M20 concrete are as follows:

- Concrete dry volume = 1.57 x 1 =1.57m^3. We have a wet volume of concrete of 1m^3 and need to convert it to a dry volume. To do so, multiply the wet volume by 1.57.

- For 1 cubic meter of concrete 8 bags of cement are required as calculated in (4.1.2.2). Rate of cement is = INR 400 per bag, therefore cost of 8 bags cement is = 8 ×400 = INR 3200.

- Sand required for 1 m^3 of concrete = 15.01 cft as calculated in (4.1.2.3); Cost of sand per cft. = INR 50. Rate of 15.01 cft sand = 15.01 x 50 = INR 750.5

- Aggregate required for 1 m^3 of concrete = 30.16 cft as calculated in (4.1.2.3); Cost of aggregate per cft. = INR 60. Rate of 30.16 cft aggregate = 30.16 x 60 = INR 1809.6

- Total cost of material = INR 3200 + INR 750.5 + INR 1809.6 = INR 5760

- In this one mason and two helpers finish 1 cubic meter of concrete work in four hours, the labor cost will be Rs. 600. Subtotal cost = INR 5760 + INR 600 = INR 6360

- Given the amount of water necessary in concreting work for mixing and curing, and assuming a 1.5% cost of water, 1.5% of 6360 = Rs.95. Subtotal cost = INR 6360 + INR 95 = INR 6455

- Overhead expenses are considered to be 2% of INR 6360 = INR 127

  Subtotal cost = INR 6455 + INR 127 = INR 6582

- The cost of equipment or tools is considered to be 1.5% of 6360 = INR 95

  Subtotal cost = INR 6582 + INR 95 = INR 6677

- Contractor profit is considered to be 10% of INR 6360 = INR 636

  Subtotal cost = INR 6677 + INR 636 = INR 7313

Therefore, the total cost for 1 cubic meter of M20 grade concrete (1:1.5:3) is INR 7313. For a further detailed overview of the cost of Cast-in-Situ technology refer to the appendix, which is included in the Delhi Schedule of Rates 2021.

4. Pre-Cast Construction Technology

Precast construction technology is a method of pouring concrete into a reusable mold or 'form' that is then treated in a controlled environment before being transported to the job site and hoisted into position. Precast Construction Technology comprises standardized and planned precast parts such as walls, beams, slabs, columns, staircases, landings, and certain unique features that are used to ensure the building's stability, longevity, and structural integrity. Design, strategic yard planning, lifting, handling, and shipping of precast pieces are all part of the precast residential building construction process. This method may be used to produce high-rise structures that can withstand seismic and wind-induced lateral stresses as well as gravity loads. The building framework is designed to achieve the largest number of mold repetitions possible. These components are manufactured in a controlled environment. The factory is built on or near the site, saving money on storage and transportation (Council, 2019).

4.1 Types of Precast Elements

(a) Precast Reinforced Concrete Element

Components including façade walls, beams, columns, slabs, stairs, and parapet walls need reinforcing bars and/or welded wire meshes to give structural strength (Table 5).

Table 5. Precast reinforced concrete elements

S. No.	Precast Components	Typical Sizes
1	Wall Panels	Sizes of panels may vary as per the requirement
2	Hollow Core Slabs	
3	Beams	
4	Staircase	
5	Columns	

(b) Molds

Steel and concrete molds for precast parts are required. Special attention should be paid to the ease of de-molding and assembly of the different sections while designing the molds for various elements. Taking into account stresses caused to the pouring of green concrete and vibration, the stiffness, strength, and water tightness of the mold are also significant (Table 6).

Table 6. Types of mold

S. No.	Mold Type	Uses
1	Conventional molds	Ribbed slabs, beams, window panels, box-type units and special elements
2	Tilting molds	Exterior wall panels where special finishes are required on one face or for sandwich panels
3	Long line prestressing beds	Double tees, ribbed slabs, piles and beams
4	Prestressing bed with an Extrusion machine	Hollow core slabs and hollow core nonload bearing wall

4.2   Machinery Used in Pre-Cast Technology

4.2.1   Hollow Core Slab Production

- Extruder Machine - EVO E120 for casting of hollow-core slabs for floor
- Extruder Machine - NANO for the casting of partition wall panels and floor slabs
- Multifunction bed cleaner machine for bed cleaning, spraying mold release agent, strand laying, etc.
- Stressing and destressing equipment, Cutting saw machine
- Four long line prestressing beds with hydraulic destressing cylinders

4.2.2   Production of Wall, Beam and Column
- Tilting Table for Wall Panels
- Magnetic Sides and Shuttering
- Beam Mold – Flexible sizing for width, thickness and length
- Column Molds – Flexible sizing for width, thickness, length and corbels.
- Stair Mold

4.2.3   Concrete Distribution
- 2 flying buckets and rails
- Discharge chutes
- Distribution buckets
- 4 EOT cranes (Electrical Overhead Travelling cranes)

4.2.4   Batching Plant
- Concrete Batching Plant - SP60 Planetary Mixer
- Concrete Batching Plant - SP60 Twin-shaft Mixer
- 5 silos for storing concrete and fly ash

4.2.5   Other Miscellaneous Machinery
- Concrete Buckets (MS)
- Bar Bending and Cutting Machine
- Power Floater
- Water Pressure Machine
- Shutter Vibrator
- Strand Decoilers
- Trampoline Laying Trolley
- Weighbridge
- Air Compressors
- DG Sets

4.2.6   Transportation and lifting Machinery
- Element Lifting- Lifting beam with hydraulic clamps.
- Lifting beam Mechanical clamps.
- EOT Crane – 10 ton (3 nos.) and 15 ton (1 no.)
- Pick and Carry Mobile Crane F160 - ACE Make
- Backhoe Loader Digimax II - ESCORTS Make
- Nylon Lifting Belts Various Capacities
- Chain and Pulley Blocks, Chain Slings, etc

4.2.7 Q.C. Machinery and Apparatuses
- Compression Testing Machine 3000 KN
- Vibrating Table
- Pan Mixture
- Oven Up to 300°c
- Weighing Scales
- Flakiness Gauge
- Elongation Gauge
- Aggregate Crushing and Impact Value Apparatus
- Sets of sieve shaker
- Measuring Flask and Jar
- Cube Molds etc.

4.3 Composition/Mix Design in Precast construction

The mix design of concrete usually used for pre-cast is M25 Grade whose design mix ratio is 1:1:2 (Cement: Sand: Aggregate) with a compressive strength of 25 N/mm^2. The following formula is used for calculating the amount of cement, sand, and aggregate necessary for 1 cubic meter of M25 concrete: (Sir, 2021)
- The wet volume of concrete = 1m^3
- mix ratio for M25 concrete = 1:1:2
- The dry volume of concrete is equal to 1× 1.54 = 1.54 m^3 (cubic meter)
- total mix proportion is equal to 1+1+2 = 4
- Where part of cement is equal to 1/4, part of sand is equal to 1/4 and part of aggregate is equal to 2/4

4.3.1 Cement Required for M25 Grade (in Cubic Meter)
Cement required for 1 cubic meter of M25 concrete in m^3 = 1/4 × 1.54 m^3 = 0.385 m^3

4.3.2 Cement Required for M25 Grade (in Kg)
density of cement = 1440 Kg/m^3
Cement required for 1 cubic meter of M25 concrete in Kgs (kilogram) = 1/4 × 1.54 m^3 × 1440 Kg/m^3 = 554 Kgs

4.3.3 Cement Bags Required for M25 Grade
1 bag cement weight = 50 kgs
Cement bags required for 1 cubic meter of M25 concrete = 554/50 = 11.08 = 11 bags

4.3.4 Sand Required for M25 Grade (in Cubic Meters)
Sand required for 1 cubic meter of M25 concrete in cubic meter = 1/4 × 1.54 = 0.385 m^3

4.3.5 Sand Required for M25 Grade (in Kg)

The density of sand is equal to 1620 Kg/m^3

Sand required for 1 cu.m of M25 concrete in kgs = 1/4 × 1.54m^3×1620 Kg/m^3 = 624 Kgs

4.3.6 Sand Required for M25 Grade (in cft)

1 cubic meter = 35.3147 cu ft

Sand required for 1 cubic meter of M25 concrete in cubic feet = 1/4 × 1.54m^3/35.3147 × 1620 Kg/m^3 = 18 cft

4.3.7 Aggregate Required for M25 Grade (in Cubic Meter)

Aggregate required for 1 cu.m. of M25 concrete in cubic meter = 2/4 × 1.54 = 0.77 m^3

4.3.8 Aggregate Required for M25 Grade (in Kg)

The density of aggregate is equal to 1550 Kg/m^3

Aggregate required for 1cubic meter of M25 concrete in Kgs = 2/4 × 1.54m^3 × 1550 Kg/m^3 = 1194 Kg

4.3.9 Aggregate Required for M25 Grade (in cft)

1 cubic meter = 35.3147 cft

Aggregate required for 1cubic meter of M25 concrete in cft = 2/4 × 1.54m^3/35.3147 × 1550 Kg/m3 = 36 cft.

Table 7. Mix quantities of 1 cubic meter of M25 grade concrete with different units

Grade of concrete	Bags	Kilograms	Cubic feet	Cubic meter
Cement quantity	11	554	-	0.385
Sand quantity	-	624	18	0.385
Aggregate quantity	-	1194	36	0.77

The setting time of precast, prestressed and post-tension concrete is 16 hours, it achieves the 28-day strength in just 16 hours because it is kept at a higher curing temperature and controlled environment (Institute, n.d.). The precast concrete element is completely ready to transport in just 2 days and after that transportation, time is something very particular and depends on location to location but on average 2 days is a standard completion time (Standard, 2011).

Table 8. Rate analysis of 1 cubic meter of M25 Grade concrete

PARTICULARS	UNIT	QTY	RATE	AMOUNT
1:1:2 (1 Cement: 1 coarse sand: 2 graded stone aggregate of 20mm nominal size) including curing				
Details of the cost of 1m³ of M25 concrete				
Materials				
Cement	Bags	11.16	350	3906
Sand	m³	0.385	1800	697.5
Coarse Aggregate	m³	0.775	1500	1162.5
Manpower				
Foreman	Each	0.05	800	40
Mason	Each	0.3	700	210
Male Mazdoor	Each	1	500	250
Female Mazdoor	Each	1	400	200
Waterman	Each	0.2	400	80
		Subtotal		6546
Tools & Tackles			1%	65.46
Power Consumption			2%	130.92
Add for sundries & contingencies			3%	196.38
Add for overhead & contractor profit			15%	981.9
		Rate per Cu.metre		7921

4.4   Cost Analysis of Precast Construction (Case Study)

The study is based on analyzing the G+1 residential model based on cost and time of construction between precast and Cast-in-Situ construction techniques. The total number of days taken by precast construction is 65, while Cast-in-Situ construction took 128 days (Figure 2). It can be seen precast is way faster than conventional construction.

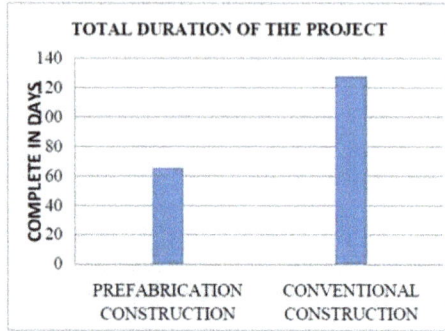

Figure 2. Total time duration of precast and Cast-in-Situ construction

The cost analysis for precast construction is based on information gathered throughout the precast manufacturing process. Because the process is the same in both circumstances, the cost of the substructure is deemed comparable. Due to the expensive initial infrastructure development cost to create the modules, the precast technique spent an extra 14 Lakhs (approximately). In the event of mass manufacture of comparable modules, this cost may be decreased (Figure 3). Due to the identical type of work, the stage finishing works exhibit a little variance of Rs. 1 Lakh approximately (Kaja & Jaiswal, 2021).

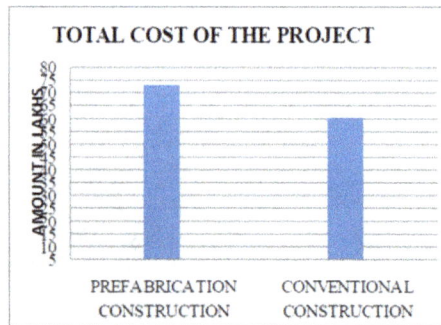

Figure 3. The total cost of precast and cast-in-situ construction

Materials Research Forum LLC
https://doi.org/10.21741/9781644902158

5.   Pre-Stressed Technology

Pre-stressing is the technique of compressing a concrete part using steel wires or strands. The tensile stresses created when the element is loaded are compensated by pre-stressing. As a result, the concrete is usually in compression (Society, 2021). The stressing of wires or cables is accomplished by attaching them to the end of a metal form that may be up to 120 meters long. Hydraulic jacks stress the wire to the appropriate tension, generally adding 10% to account for creep and other pre-stress losses. After that, the side molds are fitted, and the concrete is poured around the tensioned wires. The concrete hardens and shrinks, clutching the steel along its length and transmitting the jacks' strain to the concrete. The tensioned wires are removed from the jacks after the concrete has achieved the necessary strength. Concrete with a typical strength of 28 N/mm2 may be attained with 24-hour steam curing and the use of additives. To make shorter members, separating plates may be inserted along the length of the member and removed, allowing the wires to be shortened (Anon., 2021).

5.1   Composition/Mix Design in Pre-Stressed Construction

The mix design of concrete usually used for pre-stressed construction is M25 Grade whose design mix ratio is 1:1:2 (Cement: Sand: Aggregate) with a compressive strength of $25N/mm^2$. Refer to 4.2.2 for mix design of pre-stressed concrete.

5.2   Cost Analysis of Pre-Stressed Construction

The overall difference in the cost per cubic meter between precast and pre-stressed concrete is that there is a 10% labour cost incremental. Therefore, the total cost is equal to INR 7921 + INR 78 = INR 7999 per cubic meter. The rest of the cost remains undefined as it depends solely on transportation time and distance which can vary from location to location.

6.   Post-Tension Technology

Post-tensioned concrete is a form of prestressed concrete in which the concrete is reinforced by a tensioned reinforcing scheme. Before the concrete is poured, steel wires called post-tensioning tendons are put in plastic sleeves and positioned within the concrete formwork. The cables are pulled at each end (tensioned) and fastened to the concrete outside borders after the concrete has cured and attained adequate strength. Corrosion may occur in steel cables, as it does in any steel item. Post-tensioned concrete is so named because the tendons are tensioned after the surrounding concrete has been formed (but before the concrete structure is subjected to service loads). Instead of directly bonding with the concrete like typical reinforced concrete, post-tensioned tendons do not.

The tendons are instead encased in a protective sleeve cast into the concrete (Unknown, 2018).

The high pressures necessary to tension the tendons are passed to the concrete, causing the concrete part to be permanently compressed upon attachment. This continual compression improves the flexural and shear strength of the concrete, offering post-tensioned concrete various benefits over conventional reinforced concrete, including:

- The capacity to cover greater distances
- Shrinkage cracking resistance is improved.
- The capacity to work on soils that are softer and more expansive
- Fewer joints in the construction

6.1 Composition/Mix Design in Post-Tension Construction

The mix design of concrete usually used for post-tension is M30 Grade whose design mix ratio used here is 1:2.5: 3.5 (Cement: Sand: Aggregate) with a compressive strength of 30N/mm^2 or above after 28 days. The type of Cement used is OPC 53 Grade (Rajput, 2021).

- Specific Gravity of Cement = 3.15
- Specific Gravity of Coarse Aggregate 20mm = 2.885
- Specific Gravity of Coarse Aggregate 12.5mm = 2.857
- Specific Gravity of Fine Aggregate = 2.723
- Water absorption of Coarse Aggregate 20 mm = 0.42%
- Water absorption of Coarse Aggregate 12.5 mm = 0.47%
- Water absorption of Fine Aggregate = 1.38%

(a) Cement Content

Water Cement Ratio: 0.39

Cement content: 140.0 / 0.39 = 358.97 Kg

Minimum cement content as per IS = 360 kg/cum

Hence Cement Content: 360 Kg

(b) Mix Design Calculation

Volume of material content = material weight / (material specific gravity x total volume)

Volume of concrete = 1 Cu.M.

1 Cum = 1000 ltr (in volume)

Cement Content = 360 / 3.15 x 1000 = 0.114 Cu.M

Water Content = 140 / 1.00 x 1000 = 0.140 Cu.M.

Admixture = 1.80 / 1.17 x 1000 = 0.0015 Cu.M.

Aggregate = 1- (cement volume + water volume + admixture volume) = 1 − (0.114 + 0.140 + 0.0015) = 0.745 Cu.M.

Materials Research Forum LLC

https://doi.org/10.21741/9781644902158

(c) Volume into Weight for Concrete Mix

Material weight = material volume x percentage of total volume x material sp. gravity x total volume

Mass of coarse aggregates 20 mm = 0.745 x 0.60 x 0.50 x 2.885 x 1000 = 644.8 Kg. = 645 Kg

Mass of coarse aggregates 12.5 mm = 0.745 x 0.60 x 0.50 x 2.857 x 1000 = 638.5 kg = 639 Kg

Mass of fine aggregates = 0.745 x 0.40 x 1.00 x 2.723 x 1000 = 811.5 Kg = 812 Kg

(d) Mix Proportion per CUM. for M30 Grade of Concrete Mix Design

Cement	:	360 Kg
Water	:	140 Kg
20 mm	:	645 Kg
12.5 mm	:	639 Kg
Sand	:	812 Kg

Dosage of admixture, by the Weight of Cement

0.45% of cement weight: 1.80 Kg

Therefore, the M30 mix ratio – 1: 2.5: 3.5

6.2   Cost Analysis of Post-Tension Construction

The overall difference in the cost per cubic meter between precast and post-tension concrete is that there is a 30% labour cost incremental. Therefore, the total cost is equal to INR 7921 + INR 234 = INR 8155 per cubic meter (Standard, 2011). The rest of the cost remains undefined as it depends solely on transportation time and distance which can vary from location to location.

7.   Conclusions

This chapter analyses the different conventional construction technologies namely cast-in-situ, pre-stress, and post-tension using four primary parameters: mechanism, composition, time, and cost. The study enables the scientist and researchers an understanding of the fundamental operation of each technology, as well as the advantages and disadvantages of each, which facilitates the formulation of comparison criteria for all the building construction techniques. The chapter also paves the way for the need for an urgent transition from the conventional construction technique to sustainable building construction techniques, particularly in India.

CHAPTER 3

3-D Concrete Printing Technology

1.   Introduction

The construction industry is one of the biggest sectors worldwide, its spending is $10 trillion globally, equivalent to 13% of Gross Domestic Product (GDP) and yet it has been proven to be the least progressive as compared to other sectors. Every other sector has now or then been shifting its workflow towards being highly automated but the construction industry still relies on a manual labor-intensive workforce. 3D Printing technology also known as additive manufacturing uses a digital form of concrete that is poured out of a nozzle. This form of concrete can be tested and experimented with different admixtures to alter the strength, texture, and wastage. The concrete mix design used for printing is a slightly lower concern but the main challenge with this technology is the rebar reinforcement. Catering to this, researchers have found various other alternatives which can act as reinforcement, such as the use of Fiber Reinforced Cement Concrete (FRCC) but its strength is still a concern (Sanjayan, et al., 2019).

Concrete construction remains highly labor-intensive and accident-prone, hence automating it can help reduce both concerns. Concrete is one the most widely consumed material, its production is 30 billion tons worldwide, and hence using it with the cast-in-situ technique generates a great amount of waste and even restricts the design possibilities. In this perspective, cutting-edge building technology, novel construction materials/methods, and improved decision-making procedures can be used to claim that not only are projects becoming smarter, but they are also providing an opportunity to make our environment more sustainable. When recent building concepts and advancements are reviewed, it is clear that the sector is being shaped by a variety of construction trends. From this vantage point, it is conceivable to assert that computational models', which are now an inseparable component of a design, may control not just all stages of design but also manufacturing and management.

The 3D printing is currently one of the world's fastest growing technologies. Since the 1980s, the notion of 3D printing has progressed, although few studies have concentrated on specific 3D printing technologies. Concrete's merits as a construction material are its durability and capacity to withstand natural calamities such as rain, snow, and wind while also providing shelter. With 3D printing technology, we can work with freeform design concepts layer by layer in contrast to using a casting mold. Every printer has a limitation in terms of size like if we talk about the contour crafting method the printing zone is about 120 sq.m which corresponds to 5m x 8m x 3m and these sizes keep on varying from project to project. Therefore, the available printing zone is a significant challenge for this technology, which tends to hinder its use in large-scale projects. 3D printing technology is also being tested for immediate infrastructure construction over the moon for potential colonization purposes (Sanjayan, et al., 2019). The 3D printing is described as the method of fabricating an item from a three-dimensional model by sequentially stacking thin layers of material. The American Society for Testing and Materials (ASTM) and the International Organization for Standardization (ISO) describe additive manufacturing as the process of combining materials to create items from 3D model data, often layer by layer (Romdhane & El-Sayegh, 2020).

2. Classification Based on Material

The molten material system and binder jetting are two processes linked to Additive Manufacturing for Construction (AMC). The latter uses a concrete mix that is pumped into the nozzle (Figure 1).

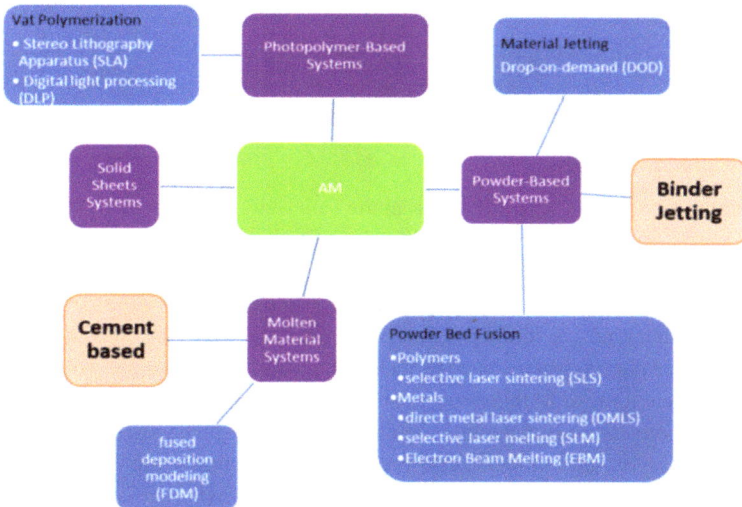

Figure 1. Types of additive manufacturing based on the material used

3. Classification Based on Technique

Various 3D concrete printing technologies have been developed in recent years to enable the use of AM in concrete buildings. These technologies are primarily based on two techniques: extrusion and powder metallurgy (Nematollahia, et al., 2017). These methodologies, as well as presently existing 3D concrete printing technologies, such as powder-based 3D concrete printing employing geopolymer, are discussed below.

3.1 Extrusion-Based Technique

The extrusion-based technology is similar to the Fused Deposition Modeling (FDM) process, which prints a structure layer by layer using cementitious material extruded from a nozzle placed on a gantry, crane, or 6-axis robotic arm. This technology has been designed for on-site building applications like large construction elements with complicated geometry (Figure 2).

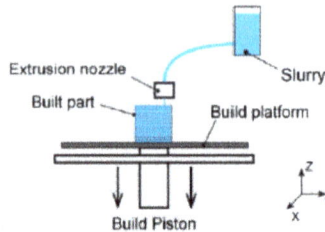

Figure 2. Extrusion-based printing schematic illustration

3.1.1 Contour Crafting

Dr. Behrokh Khoshnevis, a researcher from the University of Southern California, in the United States, developed a system called Contour Crafting (CC) in mid-2000s that paved the way for the present day's 3D Concrete Printing (3DCP). Various research institutes have commonly used Contour carving to create enormous 3D printed structures. The usage of Contour Crafting in the construction sector can reduce the amount of physical labour required for projects while also lowering construction waste. To construct a vertical concrete formwork, this method employs an extrusion-based process to extrude two layers of cementitious material. While the Contour Crafting machine is continually extruding the layers, custom-made reinforcing ties are manually put between them (at every 30 cm horizontally and 13 cm vertically). Smooth extruded surfaces are created by attaching trowel-like fins to the print head. After the extruded formwork is finished, concrete is manually poured to a height of 13 cm, and after one hour, a second batch is poured on top of the first. The purpose of a one-hour delay batch is to regulate the concrete's lateral pressure by allowing it to partly cure and harden. The improved surface smoothness and considerably increased fabrication speed are the two main benefits of the Contour Crafting technology. Another significant benefit of Contour Crafting is its ability to be integrated with other robotics technologies for the installation of interior components like pipelines, electrical conductors, and reinforcing modules to improve mechanical properties. Vertical elements are presently produced mostly in compression using the CC technique. When a doorway or window is needed, a lintel is used to span the gap, allowing the wall to be built above it. As a result, the cantilever issue is avoided (Romdhane & El-Sayegh, 2020).

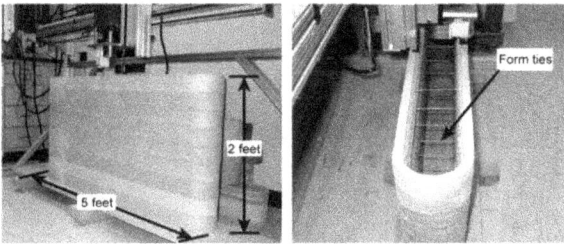

Figure 3. Concrete wall printed with custom-made reinforcement

3.1.2 Concrete Printing

This technology is similar to the Contour Crafting technology in that it employs an extrusion-based approach. Concrete Printing, on the other hand, has been created to maintain 3D flexibility with a lower deposition resolution, allowing for more control of interior and exterior geometry. Furthermore, the material employed in Concrete Printing is a high-performance fiber-reinforced fine-aggregate concrete, which results in material qualities that are superior to those achieved using Contour Crafting technology. An experiment was done using a bench that was 2 m long, 0.9 m wide at its widest point, and 0.8 m high, with 128 layers of 6 mm thickness. The bench has 12 gaps that help to reduce weight and may be used as an acoustic structure, thermal insulation, or a channel for other building functions. The bench also shows a reinforcement approach in which deliberately engineered voids serve as conduits for reinforcement once it has been installed. To make overhangs and other freeform characteristics, Concrete Printing needs extra assistance. In a similar way to the FDM process, it employs a second material. The downside of this method is that it requires an extra deposition device for the second material, which necessitates more maintenance, cleaning, and control instructions, and the secondary structure must be cleaned away in a post-processing operation (Figure 4).

Figure 4. Concrete printing at a larger scale with post-tension reinforcement

3.1.3 *CONPrint3D: Concrete On-Site 3D Printing*

While Contour Crafting and Concrete Printing technologies have numerous technical benefits, they also have certain inherent constraints, such as the need for new and sophisticated technology, tiny mineral aggregate sizes (fine-aggregate mortar rather than concrete), and the size of the printed pieces (i.e. the size of the 3D printer must be larger than the size of the element to be printed). CONPrint3D is a system that aims to provide 3D concrete printing to construction sites directly. High geometrical freedom, utilization of regularly known construction tools, and little reliance on professional personnel are the key benefits of CONPrint3D technology. One of the main goals of CONPrint3D is to not only build a time, labor, and resource-efficient advanced construction method but also to make the new process commercially feasible while gaining wider industry adoption. This is accomplished through repurposing current construction and manufacturing processes to the greatest extent feasible, as well as tailoring the new process to the limits of the building site (Nematollahia, et al., 2017). Adapting a concrete boom pump to transport material to precise areas autonomously and correctly using a custom-developed print head mounted to the boom is an important part of the project approach (Figure 5).

Figure 5. The CONPrint3D

3.2   *Powder-Based Technique*

Powder-based AM generates precise structures with complicated geometries by selectively injecting binder liquid (or "ink") into the powder bed to bind powder where it contacts the bed. This method is used to manufacture precast components off-site. The powder-based approach is ideal for small-scale construction components including panels, permanent formworks, and interior structures (Figure 6).

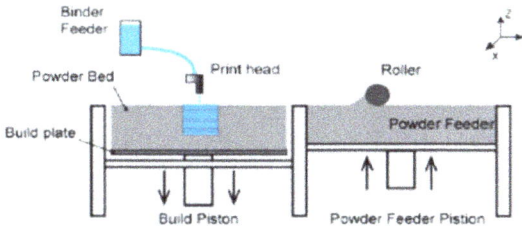

Figure 6. Powder based technique

3.2.1   *D-shape*

Enrico Dini's D-shape method employs a powder-based approach for selectively hardening a large-scale sand bed by deposition of a binding agent. The construction material and binding agents are sand and magnesium oxychloride cement (Figure 7).

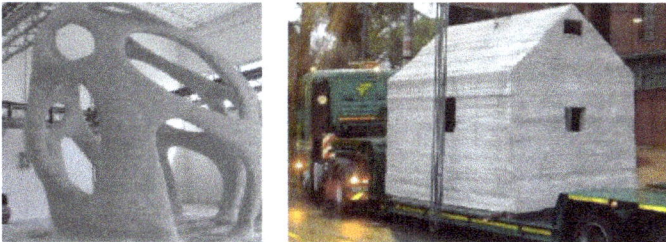

Figure 7. D-shape technique structures

3.2.2   *Emerging Objects*

By deposition of a binding agent, the Emerging Objects technology selectively hardens a unique cement composite composition. Additionally, the Shed was constructed using this method. The shed is a tiny 3D printed prototype building made entirely of Picoroco Blocks, a modular 3D printed building block for wall production made entirely of sand measuring $0.3 \times 0.3 \times 0.3$ meters (Figure 8).

Figure 8. The shed uses emerging technology

3.2.3   *Powder Based 3D Concrete Printing Using Geopolymer*

The building components having minute details and elaborate forms may be produced using a powder-based approach. In the building business, there is a need for components that can only be created using costly formwork using presently available construction technologies. To meet this industrial requirement, a powder-based approach can make strong and lasting components at an acceptable pace. The restricted range of cement-based printing materials that can be utilized in commercially available powder-based 3D printers, however, prevents this approach from reaching its full potential for usage in the construction sector. Geopolymer is a long-term replacement for OPC. It's manufactured by alkalinizing fly ash and/or slag; both are industrial by-products of coal power plants and iron production, respectively. Geopolymer has higher mechanical, chemical, and thermal characteristics than OPC and emits 80 percent less carbon (Figure 9).

Figure 9. The prototype made using Geopolymer

3D printing with significant volumes of fly ash is difficult because the strength of the material does not build quickly enough to support successive layers. Fly ash was activated with an alkali solution according to the geopolymerization technique to avoid this problem. (Figure 10) depicts a geopolymer mortar extruded via a rectangular aperture of a 4-axis gantry printer as an example of 3D printing. 5-10% slag, which is also considered one of the by-products of steel power plant sectors, was employed to speed up the reaction process.

Figure 10. The 3D printing using geopolymer

4. Mechanism of 3D Concrete Printing System

The hardware and the operating system are the two most important components of the 3D concrete printing system. The 3D printer is the system's main piece of hardware. In figure (a), a 6-axis robotic arm printer head is coupled to two peristaltic pumps (one for the premixed mixture and the other for an accelerating agent) that are installed on an industrial ABB 6620 6-axis robotic arm. The printer head and pumps are controlled by an Arduino Mega 2560 microcontroller. A gantry-based printer is distinct from the robotic-based printer. A printing head is attached to the hose that connects to the mixer pump. The printer head is mounted on a vertical arm that is controlled by a gantry system with four degrees of freedom. The end section of the printer head is a steel nozzle, and the form (round or rectangular) and size of the nozzle vary depending on the concrete printing method. Two trowels are employed in contour crafting to get a very smooth and precise printed surface, which is a significant distinction from other concrete printing procedures (Figure 10).

The operating system, or workflow, is also a critical component of 3D concrete printing. Data preparation, concrete preparation, and component printing are the three primary processes of concrete printing. A 3D CAD model is built first, then translated to an STL file and sliced with a certain layer height in the data preparation process. Finally, the printing routes for those layers are translated to a G-Code file that can be printed. Pump pressure, printing speed, and nozzle standoff distance is additional important characteristics to consider in the operating system (Figure 11). The cornerstone of a good printing process is an excellent collaboration between these parameters and the features of the printable material, such as extrudability and printability window (Yu, et al., 2017).

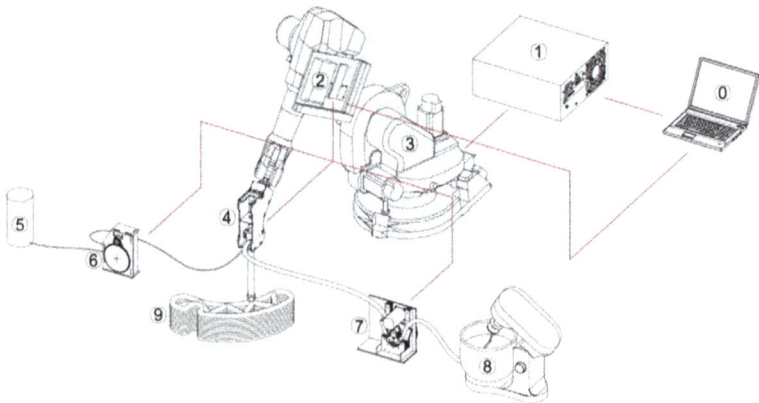

0. System command; 1. Robot controller; 2. Printing controller; 3. Robotic arm; 4. Print head; 5. Accelerating agent; 6. Peristaltic pump for accelerating agent; 7. Peristaltic pump for premix; 8. Premix mixer; 9. 3D printed object

Figure 10. Parts of a robotic 3D-printing arm

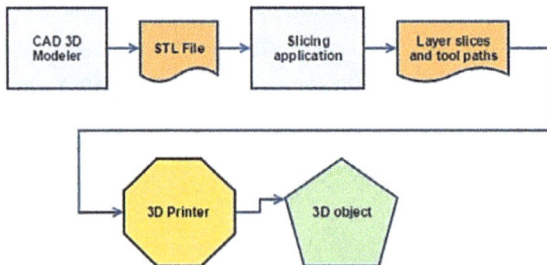

Figure 11. Procedure for 3D concrete printing

5.　Composition/Mix Design of 3D Concrete Printing

5.1　*3D-Printable Material Requirements*

The high-performance construction materials are selected because the printing method necessitates a constant, high level of material control throughout the printing. Traditional concrete cannot be utilized directly in the 3D concrete printing since no supporting formwork is needed. Low to zero slump concrete is necessary to ensure little or no distortion in the bedding layers. The low viscosity concrete, on the other hand, may be utilized for simplicity of pumping, but it will need intervention by adding a chemical accelerator to the nozzle for speedy hardening once printed (Figure 12). The granulometric qualities of the particles must be taken into consideration while making low slump concrete (Paul, et al., 2018).

Figure 12.　3D printable concrete with no slump

5.2　*Material Composition*

Several authors, who have done research related to material compositions for 3D concrete printing, are listed here. Most writers experimented with numerous different mixed compositions before settling on the perfect one. All mixes had a large number of fine particles, and there were no accounts of coarse aggregates being utilized for 3DP concrete in the literature. The maximum sand particle size was typically less than 2 mm in most circumstances. However, 3D concrete printing has also employed coarse aggregate with particle sizes up to 10 mm. Superplasticizer (SP), accelerator, and retarder were also included in the mixes to manage the open time and set duration of the materials. Water to binder ratios of 0.23-0.41, and binder/sand ratios of 0.63-0.73 were employed, according to research findings, to achieve the optimum printability and mechanical qualities of the printed material once hardened (Table 1).

Table 1. Mix design of various 3D printable concrete

Authors	Materials composition (kg/m^3)						
	Cement	Fly-ash	Silica Fume	Sand	Water	SP	Fiber
Nerrela et al.	430	170	180	1240	180	10	-
Le et al.	579	165	83	1241	232	16.5	1.2 (PP)
Anell	659	87	83	1140	228	11.6	1.2 (PP)
Perrotet al.	Cement 50%, limestone filler 25%, kaolin 25%, w/c=0.41 Binder content is expressed as a weight (SP/cement=0.3%)						
Malaeb et al.	Cement 125g, sand 80g, filler160g, water: cement=0.39, SP=0.5-1mL						
Note: Sp: superplasticier, PP: polypropylene (12/0.18 mm length/ diameter							

To maximize yield stress and form retention, three-dimensionally printable mixes are developed with a greater binder and fine aggregate content than traditional concrete mixes and SCC. It is an iterative process to create a printed concrete mix. If a concrete mix is confirmed to be compliant with the first condition, it is subsequently tested for the second. Otherwise, the mix design is altered by modifying the components, and the performance is assessed once again. This procedure is repeated until the concrete mix meets all of the printing operation's criteria. Printable concrete mixes often feature low dynamic yield stress to aid in pumping and extrusion, but strong thixotropic behavior after extrusion, to boost static yield stress, allowing the concrete to hold its weight as well as the weight of succeeding printed layers. The Table 2 below summarizes optimum printable mixes reported in the literature, demonstrating that printable mixes can be developed using various binders and fine aggregate types, as well as different water-binder ratios, sand-binder ratios, and chemical admixture, and other additives such as fibers, nanomaterial, and clays. To minimize clogging during the pumping and extrusion processes, the majority of the created printable mix designs do not include coarse aggregate. In 3D concrete printing, the eccentric screw pump is often used as an extruder, and its components, such as the rotor and stator, can only accommodate small grain sizes. The use of more binders in printed concrete mixes raises concerns about the technology's environmental friendliness, but researchers are working to reduce its carbon footprint by using eco-friendly binders and recycled materials in mix designs. The printable mix proportions are meant to be stiff and have a greater green strength so that the weight of the succeeding layers can be supported without the plastic collapsing.

Table 2. Mix design of various 3D printable concrete categories

Concrete Mix Type	Binder	Water-Binder Ration	Sand-Binder Ratio	Sand Size	Admixture (% Wob)	PP Fibres, Otherwise as stated (%Wob)
	Portland cement fly ash silica fume (0.70.0.20,0,10)	0.45	1.41	0-4.75 mm	HRWRA=0.7	-
	Portland cement fly ash silica fume (0.70.0.20,0,10)	0.32	1.5	0-2 mm	HRWRA=0.17	0.2
	OPC type II silica fume 0.90.0.10	0.43	2.3	0-2.36 mm	HRWRA=0.15 Nano-clay=03	-
	Portland cement, fly ash, silica fume (0.70.0.25.0.05)	0.35	0.75	0-1 mm	HRWRA=0.3 Clay=0.5	-
Geopolymer concrete mix	Fly ash, ,Slag, silica fume, potssium silicate,		1.5	0-2 mm	Nano-clay=1.2 Fiber=0.25	-
	Fly ash, slag (0.50.0.50)	0.36	1.5	-	Retarder=0.5 Alkali activator=10	-
Fibre-reinforced composite	Portland cement, flyash, silica fume (0.70.0.20.0.10)	0.26	1.19	Average size= 0.39 mm	HRWRA=1.8	Basalt fibers, 0.5
	Portland cement, silica fume (0.70.0.30)	0.16	1	-	HRWRA=1.5 VMA=0.1	Steel fibers 2% by voulme
Engineered cementious composite (ECC)	Portland cement, sulfoaluminate cement, flyash (0.40.0.03.0.57)	0.28	0.40	0-0.3 mm	HRWRA=1.2 VMA=0.1	Polyethylene fiber 2% by volume
	Portland cement, calcium aluminate cement, Fly ash (0.30.0.02.0.68)	0.25	0.38	-	Nano-clay=0.3 VMA=0.3 HRWRA=0.9 Nano-TiO2=5	PVA fiber 2%
Underwater concrete	Portland cement and limstone (0.65.0.35)	0.38	1	0-2 mm	HRWRA= 0.5%, 1%, 1.5%, 3% Anti-wash agent= 0.5, 1.1, 1.5	-

Cement paste-based ink	Calcium sulfoaluminate cement, metakaolin (0.97.0.03)	0.35	Cement paste	Superplasticizer=0.3 VMA=0.4 Retarder=0.15	-
	Cement type II, silica fume (0.975.0.025)	0.3	Cement paste	1.5	-
	Portland cement	0.26	Cement paste	HRWRA=0.4 VMA=1.2	-
HRWRA= high-range water reducing agent, VMA= viscosit modifying agent, PP= polypropylene fibers, WOB= weight of binder					

6. Requirements of Concrete for 3D Printing

With the passage of time, the use of concrete and construction materials together with 3D printing continues to increase. The method, which has started with small, non-structural uses, has begun in this context with the growth of large enterprises large structures adopting and developing the 3D printing method. Figure13 shows the step-wise process of preparing a 3D printable concrete.

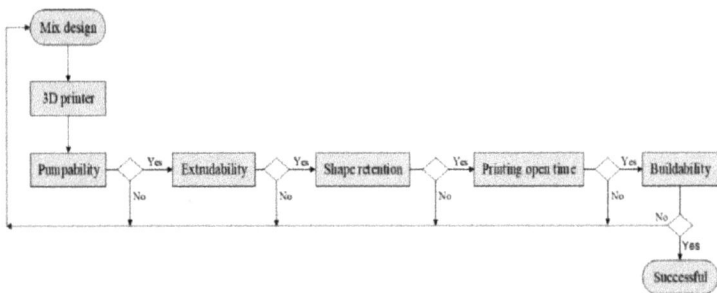

Figure13. Step-wise process of preparing a 3D printable concrete

To excel in 3D printing concrete structures, the following key properties of concrete require improvement:

6.1 *Extrudability*

Extrusion mechanism is the ability of a concrete pump, transmission tube and spray nozzle to move through without any alteration in its physical properties. The co-application of self-compacting concrete and shotcrete concepts for the mixture model achieves strong extrudability (Le et al. 2012(a)). The concrete must have a fluid rate so that the printer can be fed into layers. If there are many architectural specifics in the

Materials Research Forum LLC
https://doi.org/10.21741/9781644902158

design, the pressure level must therefore be adjusted so that the tool does not waste the extrusion material. Excessive concrete construction leads to poor coating surface area during printing resulting from incorrect control of the system (Nadarajah2018). Many studies mention the concrete layers with extrusion and the fresh properties of 3D concrete, but a suitable test for evaluating the concrete property has not been defined.

6.2 *Buildability*

The constructiveness can be described as the capacity of the concrete layer underneath to hardness and hold other layers before putting the next concrete layer on the printed surface. In this way, an adequate base for building the concrete on each floor is given. The most important elements of 3D concrete is extrudable and capable of construction both these criteria contribute to the concrete workability (Le et al. (2012) (a)). Bos et al. (2016) found that the sub-layers should not be deformed by the weight of the top layer in the 3D production method, but that the interlayers should also bind and be well attached to the top layers to be able to shape. Chemical admixtures, temperature and use of less gypsum cement are factors that affect buildability.

6.3 *Workability*

The quality of the finished printed design is greatly affected by the fresh concrete quality once the concrete is poured, stays intact and has enough working capacity (can be extruded). Small changes in environmental conditions (temperature, humidity, raw material moisture, etc.) influence the workability of 3D printing concrete (Papachristoforou et al., 2018). In order to increase the workability of the modifying agent of the 3D concrete mixture viscosity, it must have a small particle size to match the diameter of the nozzle. The modification of the pumpability and extrudability workability of 3D concrete is close to the Shotcrete Concrete wet system development (Lim et al., 2011). The use of viscosity-modifying agents in 3D concrete printing is important, because the rheology is modified by applying thixotropic properties to the concrete. Thus, the viscosity of the concrete decreases when the strength is applied and when power is stopped, it increases the viscosity so that the concrete is given a good speed resistance (Ozalp et al., 2018).

6.4 *Open Time*

A cement material's workability time is usually associated with setting time, determined by a vicat. This device is designed to determine the start and end times and cannot be used over time for determining changes in the workability of fresh concrete. Several experiments were conducted using the crash test to track the change in functionality over time. Nevertheless, a crash test to determine the open period is not acceptable. In terms of calculating workability, calculation of sliding force overtime produces more insightful

3D Concrete Printing Technology
Materials Research Foundations **134** (2022)   https://doi.org/10.21741/9781644902158

results. The Open-time period is established as the time-frame for maintaining exclusivity by working the fresh concrete for 3D concretes (Le et al., 2012 (a)). This ensures that the open time is when the pump, printability and buildability of the 3D concrete are compatible within reasonable tolerances.

6.5   *Contact Strength between Layers*

The software to get a solid structure adherence must be strong when placing the concrete on top of each other. Therefore, the concrete should not be hardened, but instead of hydrating the concrete, the preceding layer should continue when you put the concrete on the ground. In other words, there must be no cold joint. Many researchers have emphasized the importance of the form of the layers to ensure this. It is possible to change the shape of layers by varying the shapes of the printing nozzle (Bos et al., 2016). The printer will rotate around the corners, apart from the size and shape of the dots, to correctly match the corners to the structure. The 3D concrete printer's durability is checked many times before actual printing to provide corners of 90° (Wolfs, 2015).

6.7   *Aggregates*

Aggregates play a crucial role during the 3D concrete phase of the aggregates determines the structure's load-bearing strength. Figure 14 shows the comparison of material percentage by volume in CC, SCC, and 3DPC. The size of the nozzle varies from 20 mm to 40 mm. Hence, the width of the aggregates should, be greater than 4-6 mm to avoid obstruction of the nozzle. Additionally, the use of coarse aggregates contributes to instability in the printing system, allowing the framework formation to collapse (Nadarajah, 2018).

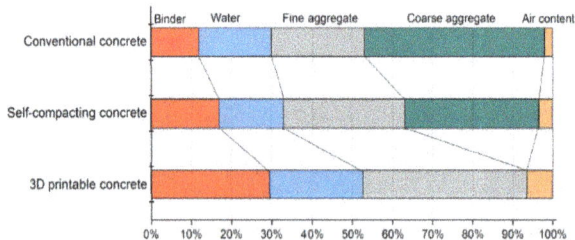

Figure 14. Comparison of material percentage by volume in CC, SCC, and 3DPC

6.8   *Water Cement Ratio*

Many scientists have experimented with the water-cement ratios between 0.25-0.44. It is necessary to use the minimum amount of water with superplasticizers for better concrete adhesion. Furthermore, it was determined that adding materials such as fly ash, silica fume and slag could benefit the mixture if 5 to 30 percent of the total binder volume were added (Nadarajah, 2018). As a result, the water-binder ratio is lower and the binder-sand ratio is larger. Chemical admixtures such as viscosity altering admixtures, superplasticizers, accelerators, and nanomaterials are used to further tailor the properties of concrete ink for printability and vertical constructability. The most often recorded water-binder ratio is between 0.30 and 0.40, while the most common sand-binder ratio is between 1.2 and 2.0. Researchers have also produced printed concrete reinforcement mixtures with high fiber content. Additional study is required to optimize fiber-reinforced printed concrete mixes such as engineered cementitious composites and ultra-high-performance fiber-reinforced concrete (Rehman & Kim, 2021).

6.   Challenges for 3D Printable Material in Large Scale Construction

The 3D printing technology needs to be developed for mechanical strength, strengthening, curing and durability. The performance of 3D concrete printing is studied in some respects. For 3D concrete printing, specifically on in-situ applications, the manufacturing conditions should be investigated. The 3D concrete has some feasible features such as flowability. These characteristics vary from traditional concrete. Therefore, the development of 3D concrete should be performed with greater care by taking these characteristics into account. Obtaining printable cement materials compliant with 3D printers is one of the fundamental challenges. Because of insufficient rheological and rigid properties, high-performance and highly durable cement-based products currently available cannot be treated directly in a printing process. The rheological properties of cement, setting time and temperature of hydration should be appropriate with 3D concrete open time. There is, however, no standard for 3D concrete printing technology around the world. To test the mechanical behavior of samples, components and structures effectively and reliably made with cement material through 3D printing, fundamental and unified standards must be established.

The success of 3D Printing is dependent on the material's qualities, notably the fresh features of 3D printing materials including workability, viscosity, and green strength. No-slump yet pumpable material is typically required for 3DCP to avoid deformation/settlement of bead layers (printing layers). Another option is soft material, which allows for slight slump but returns to its original hardness once the material is put via the nozzle head. Figure 15 depicts the needed viscosity profile of any suitable 3D printed material (Roussel, 2018). It can be seen that material must have a higher viscosity before extrusion (i.e., before deposition via the nozzle head), which then decreases during the printing phase. The material can then be smoothly pushed from the mixing machine to the hose pipe and deposited through the nozzle head. The substance must restore its previous viscosity after being deposited so that the next layer can be deposited over it. The distortion of the bead layers can be adjusted in this way. This is difficult because it necessitates careful material composition selection, which may not be a key factor in traditional concrete. The hose pipe and nozzle head are not required for most typical concrete materials. Pumping concrete, also known as self-compacting concrete (SCC), is frequently employed in existing concrete construction projects because of its advantages in terms of time savings and the absence of vibration. SC's increased slump flow, on the other hand, may not be suitable for 3D printing.

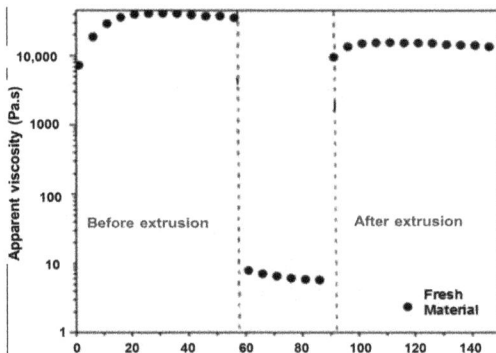

Figure 15. Required viscosity profile of fresh 3D printable material.

7. Advantages of 3D Concrete Printing

Following are the advantages of 3D concrete printing:

- Using 3D printing technology to create structures would significantly lower the cost associated with it since it would minimize the number of materials utilized and also eliminate material waste. Additionally, it decreases labor costs since the required workforce is reduced, hence lowering the cost of accidents and injuries.

- The designs submitted in the form of CAD files may be printed immediately. When a building project is completed with a 3D printer, the construction time is drastically reduced. Thus, 3D printers can manufacture a single-story building in 12 to 24 hours, allowing for rapid structural development. With technological improvements, it has been shown that buildings produced utilizing 3D printings have increased strength and endurance and can survive temperature fluctuations.

- Even elaborate designs may be precisely produced without incurring additional costs and allowing for more efficient use of space (Wolfs et al., 2018).

- This approach is more environmentally friendly since the majority of materials utilized in these methods are recyclable or reused. Additionally, this process has a far less carbon impact than traditional procedures. The printed construction will only generate roughly 30% of the garbage that a typical construction project produces while utilizing very little energy (Wolfs et al., 2018).

- Since there is savings in raw materials and, more crucially, the labour cost, the 3D printed buildings are much less expensive to construct than those constructed using traditional methods. By having most of the building performed by 3D printers, an architectural project's labour expenses can be lowered by up to 80%.

- This process facilitates the verification of designs by generating prototypes, which eliminates the possibility of mistakes occurring during mass manufacturing. The need for formwork, scaffolding, and other supporting structures may be omitted, saving about 40% of the building cost.

8. Disadvantages of 3D Concrete Printing

Following are the advantages of 3D concrete printing:

- There are no rules or procedures regarding building regulations for obtaining approval for 3D printed structures for home or commercial usage. First, the government would have to establish electrical, plumbing, structural integrity, and public safety rules that must be obeyed.

- Concrete and polymers are about the only materials that can be delivered from the printing head.

- While 3D printing may minimize the cost of building, the printers themselves are rather expensive, which is a disadvantage.

- The cost of 3D printing equipment and materials makes the technology prohibitively expensive. Industrial 3D printers are still quite expensive, costing hundreds of thousands of dollars, making the technology's initial costs very high. Capital investment for a single machine starts in the tens of thousands of dollars and can reach hundreds of thousands of dollars (Wolfs et al., 2018).

- 3D printers have come across as easy to operate and sound more useful than they really are due to the excitement and possibilities surrounding (Wolfs et al., 2018).

3D printing technology. Skilled workers proficient in 3D printing and CAD are needed.

- 3D printers are energy-intensive. According to the study, it uses around 100 times the amount of electrical energy as traditional procedures.

- Because the printers are so large, they create storage issues on-site.

9.   Conclusions

The 3D printing technology is one of the techniques that can extensively change the future of the industry; it is an upcoming automation technique that uses layer-by-layer deposition of material to build a structure. Talking about the conventional cast in situ construction, 60 % of the cost of construction is only for the formwork made out of timber, which sooner or later is going to be discarded because it has a tendency to be reused for only a certain number of times. Therefore, this creates an urging demand to shift our methods towards more sustainable and greener ways (Sanjayan, et al., 2019). Today concrete alone is a well and extensively researched material; its properties can be tweaked by adding different types of admixtures such as adding fiber reinforcement along with rebars can increase the tensile strength of the concrete. Adding water reductants can help consume less water, therefore, reducing water usage, and accelerating admixtures help to decrease the setting time of concrete cast which eventually decreases the time of construction.

CHAPTER 4

Properties and Cost Analysis of 3D Concrete Printing

1. Introduction

Additive manufacturing is a recent revolution in the construction field since cementitious materials became printable. This extrusion technique has enabled the construction of very complex geometry with a reduction in costs, time and labor interventions. Although different factors affect the properties and printability of the concrete mixtures, the mixture composition has a great influence on the final properties of 3D printed concrete. For instance, the size of aggregates is an influential factor in the final strength of printed concrete but there are considerations to use up to a specific size allowing the printing process. Mostly fine particles are used for many different mixes of concrete when 3D printed. This limitation is associated with the size of the nozzle and the capability of the machine to continuously extrude. The maximum size of sand particles that have been used in the majority of studies is not more than 2 mm, but in some cases, aggregates up to the size of 10 mm are used (Rushing et al., 2019). The concrete's shear stress, viscosity, open time, and green strength are vital for pumpability, extrudability, buildability, and interlayer adhesion. The concrete's open time refers to the change in flowability over time, allowing printing without affecting print quality or hardened qualities.

2. Rheological Properties of 3D printable materials

The different rheological properties of 3D printable materials are the pumpability of concrete, the extrudability of concrete, and the buildability of concrete. They are summarized as below.

2.1 Pumpability of concrete

The material's mobility and stability under pressure while keeping its basic qualities are referred to as pumpability. A moderately soft material that is simple to pump is needed at the pump, whereas a rigid material that does not droop or lose form is desired at the nozzle. The concrete is a heterogeneous material with a wide range of particle sizes, shapes, and densities. As a result, a well-designed and optimized mixture design is essential to provide optimum pumpability. There are two ways to improve pumpability: by adding enough paste to establish a thin smearing border layer over particles, and ii) by choosing the right grout consistency and structure between aggregate grains to prevent forced or pressured bleeding during high-pressure pumping. It is important to have an appropriate consistency, and resistance against water squeezing out of the concrete owing to high pressure in the pipe, a process is known as pressured bleeding (Paul, et al., 2018).

2.2   *Extrudability of concrete*

A very viscous, plastic-like material is pushed through a die/nozzle, which is a rigid orifice with the desired cross-section, in conventional concrete extrusion, for example, to manufacture pre-cast hollow-core slab parts or pavement curbs. The materials are created under intense shear and compressive stresses in this manufacturing process. This is altered in 3D concrete printing by the necessity of pumping new material across suitable distances based on the size and working volume (length, breadth, and height) of the printer (gantry, robot, or crane). To be pumpable, thixotropic concrete material is required, as explained in the preceding section. The new concrete reaching the nozzle is less viscous than in conventional extrusion because of the lower viscosity induced by the pumping pressure. Shape retention requires a robust extrude capable of quickly recreating its geometrical shape by re-flocculation. Because the nozzle is smaller than the pipe cross-section, there may be an increase in pressure at the nozzle. Segregation must also be avoided in this case. Material constituents and amounts should be carefully selected and managed to obtain the desired thixotropic and tribological (pumpability) properties. Segregation may result in obstruction in the pipe due to inadequate matrix composition.

2.3   *Buildability of concrete*

The capacity of the bedding layers to retain the succeeding layers on top of them without collapsing or deforming is referred to as buildability, which combines the traits of printed layers being self-supportive with shape preservation. When consecutive layers are applied, layer imperfection and settling may cause instability. This is not a problem in conventional concrete buildings. Since 3D concrete printing does not need any formwork, it must be self-supporting. Changing the nozzle type is a simple technique to enhance structural buildability. A circular nozzle orifice has a smaller contact area between the two beads than a rectangular or square aperture, as illustrated in Figure 1(a). Increasing the number of nearby filament layers, as seen in Figure 1(b), is another option to increase buildability. Note that in 3D concrete printing, the word filament refers to a printed layer of concrete that resembles a fine thread. Thin 3D concrete printing objects may not be suited for this strategy. Research showed a considerable increase in buildability by increasing the number of neighboring filament layers from one to six, as illustrated in Figure 1(b), for the same material composition and nozzle aperture. A supporting filament, as illustrated in Figure 1(c), may be employed to increase the buildability by forming a cellular-type structure. The printing loop must be closed for printing two adjacent layers with a single nozzle head, i.e. the start and finish of printing must be at the same location. However, with multiple nozzle heads, these two layers may be produced at the same time (Paul, et al., 2018).

Figure 1. Influential parameters on 3D concrete printing buildability, a) nozzle orifice shape, b) number of filament layers, and c) supporting filament in the gaps

3. Mechanical properties of 3D printed vs casted steel fiber reinforced concrete

To determine the properties of a different component, experimentation conducted shows that the specimens retrieved from the 3D concrete printing specimens had lower compressive strengths than the cast specimens. The reduced air content, segregation by printing of non-optimized material, and curing conditions are some of the factors that might have contributed to these contradictory findings. The strength qualities of the printed specimen are affected by the loading direction. It was discovered that while testing parallel to the layer depositions, the compressive strength was greater than when testing perpendicular to the layer depositions. This is a difficult thing to describe. It's worth noting that their specimens were heat cured and examined at only 3 hours young. The failure mechanism of printed cubes subjected to compression testing was also looked at. The failure patterns of specimens loaded in the X, Y, and Z directions, as indicated in the Figure 2, were virtually identical. The examples all failed diagonally, with two sets of triangular fractures meeting at the center and forming an hourglass shape on opposing sides (Paul, et al., 2018).

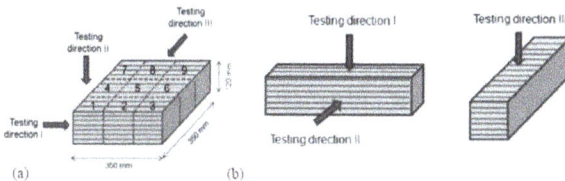

Figure 2. Specimen collection for 3D concrete printing and testing direction

4. Types of Reinforcement Strategies for 3D Concrete Printing

The various types of reinforcement techniques for 3D concrete printing are:

4.1 Cable Introduction at the Nozzle

At the nozzle, a continuous reinforcing cable is introduced into the extruding concrete. The print head is coupled to a reel on which a reinforcing cable is wrapped, and the open end of the reinforcing cable is inserted into the nozzle via a hole. Simultaneous extrusion of concrete and reinforcing cable via the nozzle winds reinforcement cable from the print head's spool and inserts it into the concrete filament in a continuous stream. The cable's binding strength with printed concrete was weaker than its bond with casted concrete, according to the pull-out test findings. In a four-point bending test, cable-reinforced printed concrete revealed ductile failure behavior with cable slip. The flexural response of the printed concrete may be enhanced using this reinforcing approach, according to bending test findings. Although the addition of a reinforcing cable at the nozzle improves the flexural strength and ductility of printed concrete, the interface between the reinforcing cable and the printed concrete has been observed to be porous, necessitating additional research (Rehman & Kim, 2021).

4.2 Insertion of Reinforcing Elements into the Printed Concrete

Another option is to print a concrete layer first, and then use a device connected to the print head to inject reinforcing pieces into the produced concrete. Several researchers have reinforced printed concrete by introducing steel nails in various orientations, resulting in greater flexural reinforcement. Another method that has been tried is to inject steel fibers perpendicular to the printed layer interface. The addition of fibers to printed concrete improved its flexural strength and ductility. Other researchers recommended stapling extruded layers with U-shaped reinforcing wires at the same time. Although these ways of reinforcing concrete can improve its mechanical qualities, the majority of them are manual. The development of automated print heads for putting reinforcing components into extruded concrete requires research (Rehman & Kim, 2021).

4.3 Mesh Reinforcement

Sanjayan studied reinforced concrete by using a custom-designed nozzle to insert vertical steel mesh in the extruding concrete filament. Mesh reinforcements were layered and overlapped to provide a continuous reinforcement over the height of the printing wall. The experiments revealed that the mesh and concrete had a strong connection. The use of embedded mesh reinforcement resulted in a 170–290 % improvement in flexural strength. The polymeric mesh was also employed to reinforce concrete by the researchers.

Interlayer polymeric mesh improved the ductility of printed concrete in compression testing (Sanjayan et al., 2019).

4.4 Printing over Conventional Bars

The horizontal positioning of reinforcing bars on newly printed layers is followed by the printing of further layers over them in this procedure. The impact of concrete workability on the bond strength between steel and printed concrete was studied by printing concrete over a standard reinforcing steel bar (8 mm). A high thixotropic printable mix was found to generate a satisfactory bind between concrete and steel. In these investigations, the placing of bars has been done manually, but it can be automated (Rehman & Kim, 2021).

4.5 Use of Printed Reinforcement

Another scientist Mechtcherine suggested using a gas metal arc welding technology to 3D print steel reinforcement. The ductility and bond strength of printed steel bars were found to be comparable to those of conventional steel bars. The researchers used 3D printing to create plastic formwork with ribbed structures to replace steel reinforcement in 3D-printed concrete. After setting, concrete forms a composite structure with printed plastic formwork, considerably increasing its flexural strength. Adopting a method based on printed reinforcement would need the employment of two distinct setups, one for printing reinforcement and another for extruding concrete, which might increase operating expenses (Rehman & Kim, 2021).

4.6 Fiber-Reinforced Printable Concrete Mix

The fiber-reinforced concrete and engineered cementitious composites (ECC) are being studied as alternatives to the aforementioned reinforcing technologies in 3D concrete printing. At the extruding nozzle, such cementitious compositions do not need any further automated setup. Extruding fiber-containing concrete improves mechanical performance by aligning fibers parallel to the printing direction. Researchers examined the effects of 0.25 percent and 1% short glass fibers in geopolymer concrete on compressive, flexural, and tensile strengths. The inclusion of fibers greatly enhanced the flexural and tensile strengths, according to test data (Figure 3).

Natural fibers are both environmentally beneficial and less expensive than synthetic fibers. These environmentally friendly chemicals might be employed as reinforcing agents in printed concrete. The effects of natural fibers on the rheology and printability of concrete, as well as the structural performance of printed concrete, will need more investigation.

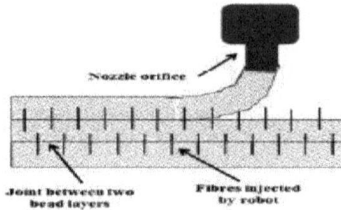

Figure 3. Fiber reinforced printable concrete

4.7   Post-Printed Reinforcement Strategies

The concrete structure is initially printed in different planned segments on a flat surface and then reinforced using post-processing procedures such as segment assembly, reinforcing bar passing through holes, post-tensioning, and grouting. Although this approach seems to be more viable for big load-bearing structures, post-processing stages need labor (Rehman & Kim, 2021).

5.   Printability Window

A concrete mixture's printability window is a property. Given the workability loss that happens over time, this window is the amount of time during which the printing mixture might be extruded by the nozzle with acceptable quality. The supply of material to the nozzle at the right time is critical to the functioning of a full-size construction printer like the CC. The printability window of a combination, as well as the printability and blockage time limitations, are defined in this research using two-time restrictions. The printability limit is the time when workability loss affects the quality of the printed layer, as defined by the triple print quality requirements, whereas the blockage limit is the time when the concrete cannot be guided out of the printing nozzle at all, resulting in mixture solidification and nozzle damage if the process is prolonged. The first setting time of concrete is an essential related subject. The first setting time is defined as the time necessary after cement water contact for the mortar sieved from the concrete to attain a penetration resistance of 3.5 MPa, according to ASTM C125. The first setting time of created mixes was measured using a concrete penetrometer in this investigation. Three further combinations were generated and investigated since the setting durations of the four mixtures were comparable around 325-350 minutes (Sanjayan, et al., 2019).

Calcium chloride ($CaCl_2$) is often used as a cementitious mixture accelerator because it works as a catalyst for C_3S processes, resulting in increased hydration heat liberation. As a consequence, three different doses of analytical grade $CaCl_2$ were added to the PPM mixture (1 percent, 2 percent, and 3 percent of Portland cement mass), and the resultant mixes were called PPM1% CaCl, PPM2% CaCl, and PPM3% CaCl, respectively. The concrete penetrometer readings were repeated three times for each combination. The initial setup times of PPM, PPM1% CaCl, PPM2% CaCl, and PPM3%CaCl were 335, 237, 181, and 163 minutes, respectively, based on penetrometer data. $CaCl_2$ increased the hydration process and shortened the setting time, as expected. The Table 1 shows the results for printability window settings as well as the initial setup time. Beginning 20 minutes after the first water-cement contact, a single layer was printed every 5 minutes to establish the printability limit. The printability limit was defined as the earliest period when print quality standards were not met. Similarly, the blockage limit was defined as the earliest period when concrete could not be extruded from the nozzle. The findings show that nozzle blockage may occur long before the mixture's first setting time and that setting time cannot be utilized as an alternative signal. During construction, nozzle obstruction may result in severe time loss, nozzle damage, and additional costs. As a result, evaluating the blockage limit for each combination during mixture design and laboratory testing is advised. Setting time and workability loss measures, similar to those used in conventional concrete, might be employed to investigate the impact of various chemical admixtures on the fresh-state behavior of printing concrete (Sanjayan, et al., 2019). Hence the printability window plays a major role in large-scale construction projects because the concrete needs to be laid down before it gets semi-hardened. This is a challenge in 3D concrete printing where the printability window and blockage limit extents need to be carefully analyzed.

Table 1. Initial setting time, printability limit and blockage limit of mixtures (min)

Mixture ID	Initial Setting Time	Printability Limit	Blockage Limit
PPM	335	55	85
PPM 1% CaCl	237	40	75
PPM 2% CaCl	181	40	60
PPM 3% CaCl	163	45	44

6. Cost Analysis of 3D Concrete Printing

The construction cost estimation of 3D printing cannot adhere to the principles of the traditional building since it has been classified as both an industrial and a construction product. To understand the cost of 3D concrete printing various factors of onsite and offsite construction need to be considered. The major cost in 3D concrete printing is the skilled labor and material (Table 2).

Table 2. Cost calculation of printed components

No.	Cost composition	Cost calculation
1.	Factory price Cost of 3D printing Manufacturing cost Management fee Profit	= (a) + (b) + (c) + (d) = Direct labor cost and material cost = Detailed design cost + electricity and water charges for printing + depreciation of equipment + intangible amortization + salaries + labor protection expense + environment protection fee + seasonal cost + loss during machine maintenance. = management fee + accounting cost + sales cost Determined by manufacturers
2.	Transportation cost	Transportation distance x unit price
3.	VAT & The final cost of printed components	According to local regulations = (1) + (2) + (3)

6.1 Parameter Selection and Strength Requirements of Geopolymer Printable Concrete

The strength of the geopolymer is determined by several parameters, including the curing temperature, chemical concentration, solid solution ratio, curing time, alkali solution molar ratio, and the concentration of elementary materials containing silica and aluminum, type of additives and admixtures, and alkali solution type. M30 (concrete standard grade) is used to determine the influence of mix and particle size on the compressive strength of a geopolymer. M30 concrete is regarded as an excellent choice for durability and exposure to harsh environments. Table 3 calculates the mix percentage, necessary material, coarse aggregate size, and unit price required to manufacture 1 m^3 of geopolymer concrete (Munir & Kärki, 2021).

Table 3. Cost production [EUR] of 1 [m3] geopolymer concrete

Concrete Grade	Mix Proportion	Materials	Course Aggregate (mm)	Price (Euro)	Unit	Quantity (kg)	Amount (Euro)	Total Price (Euro)
M30 (Concrete standard grade)	Design Mix	Ash	2-4	156.8	t	350	54.88	860.07
		C & DW	<4	416.4	t	610	254	
		Green liquor sludge	-	75.5	t	262	20	
		Flotation sand	2-4	59.1	t	583	34.5	
		Fiber rejected	-	73.2	t	420	30.8	
		Sodium silicate	-	4	L	116.36	465.44	
		Water	-	10	M^3	45	0.45	

6.2 *Analysis of 3D Concrete Printing Cost Using a Case Study*

The Apis Cor manufactured home had a 38 m2 size, but the cost of a 3D printed house is calculated using a 50 m^2 space in this research. Table 3 displays the anticipated cost of constructing a 3D-printed home on the job site. Hours worked, labor costs, material costs, and printing time are all variables. The cost of acquiring 3D printers is not included in this estimate of the cost of a 3D-printed home. The Apis Cor estimates that the requisite partitions, self-bearing walls, and building envelope will take less than one day to construct, with a total cost estimate of EUR 8330.94 (Figure 4).

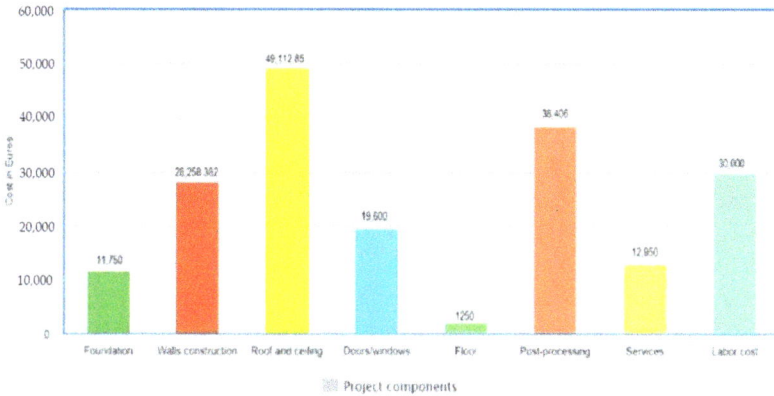

Figure 4. 3D printed house cost estimation

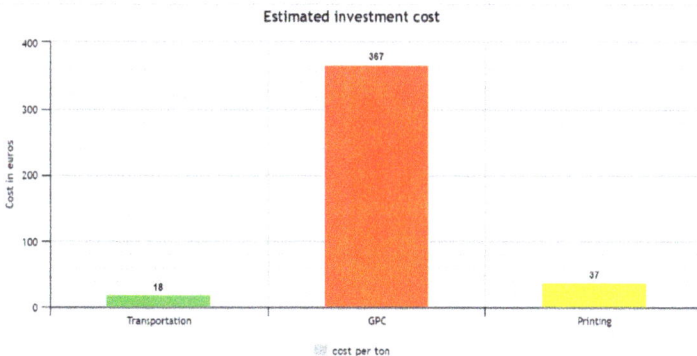

Figure 5. Investment cost needed per ton for each stage

The most essential step in geopolymer 3D printing cost calculation is raw materials pre-treatment processes, which has a substantial impact on the entire output cost. Pre-treatment of raw materials accounts for around 87 percent of the total cost of the finished product (Figure 5). GPC's strength comes from the waste material utilized, which offers long-term sustainability and uses less energy than OPC manufacture. M30 GPC has a little higher manufacturing cost than OPC of the same grade. As a result, GPC 3D-printed buildings are somewhat more expensive than OPC-printed houses (Munir & Kärki, 2021).

Table 4. Cost comparison between Conventional and 3D printed house

Construction Method	Area m^2	Average Price (in Euro)
Traditional Technology and OPC	50 m^2	280,000
Traditional Technology using GPC	50 m^2	284,760
3D printed house using OPC	50 m^2	189,000
3D printed house using GPC	50 m^2	192,327

7.   Conclusions

In Figure 6, it can be observed that the cost of concrete with 3D concrete printing is much higher as compared to other conventional technologies, this is because the initial cost of the material preparation is higher. This technology cut the labor, formwork, and other equipment costs due to which eventually the cost of a particular project comes at par with the conventional techniques as seen in Table 4. Considering the time of construction then 3D concrete printing is surely the winner because the initial setting time of this is just 5.58 hours and one can finish a small project in just 12-13 weeks. The reduced time has reduced the cost which neutralizes the extremely high cost of geopolymer concrete. The other technologies other than cast-in-situ although attaining full strength in just 16 hrs require additional curing, reinforcement, and transportation to the site due to which the time of construction increases, on the other hand in 3D concrete printing the printer prints a floor continuously in no time and finishes the project (Figure 7). The concrete used here is geopolymer concrete which is a form of green concrete using fly ash or slag as one of the components hence it requires some form of pre-treatment which increases the cost. This concrete helps in achieving higher strength as compared to conventional digital concrete. Also, using 3D concrete printing in large-scale projects can cut down the cost to a certain extinct because it saves installation, labor and many other cost factors. So, the 3D concrete printing is an efficient technique among all five types of technologies, although more research is required in this field for the future to implement it at a larger scale.

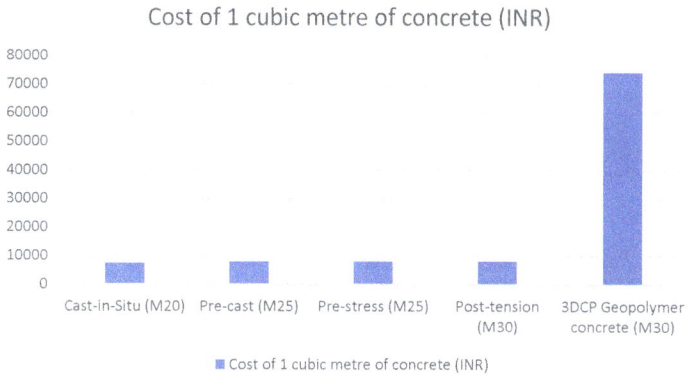

Figure 6. Cost of 1 cubic meter of concrete of different technologies

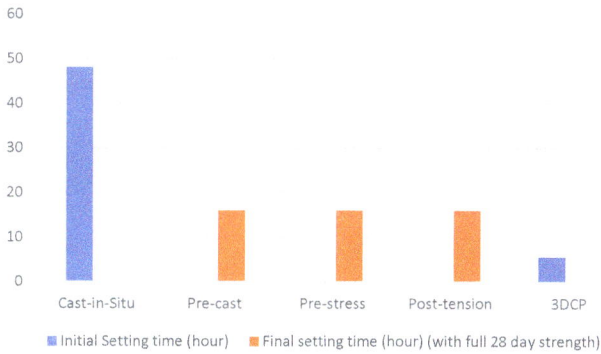

Figure 7. Setting time of different technologies

CHAPTER 5

Green Concrete

1. Introduction

Green concrete is simply a form of concrete that is made out of eco-friendly waste materials. 'It utilizes at least one waste material such as recycled demolition waste aggregate, recycled concrete aggregate, blast furnace slag, manufactured sand, glass aggregate and fly ash'. The concept lies in creating a type of concrete that uses as many recycled materials and produces the least amount of carbon footprint (Agarwal & Garg, 2018). Green concrete is a word used to describe concrete that has gone through extra procedures in the mix design and installation to provide a long-lasting construction with little maintenance. It should also be high-performing and long-lasting over its whole life cycle. Green concrete, in other terms, is a form of concrete that is favorable to the environment. Green concrete helps to improve the three pillars of sustainability: environmental, economic, and social consequences. The amount of Portland cement replacement materials, manufacturing process and procedures, performance, and life cycle sustainability impacts are all considered to determine whether the concrete is green. Green concrete is inexpensive to create since it is made from waste materials, reducing energy usage while increasing strength and longevity. To make green concrete, many properties such as mechanical properties, fire resistance, durability, strength, thermodynamic properties, environmental properties, and so on are taken into account (Goyal & Kumar, 2018).

The construction industry faces new challenges at every stage of advancement, while the field requires advancement toward automated construction, on the other hand, also requires a structural eco-friendly material to support that automation era. By volume, concrete is the most consumed material worldwide, and its production gives out CO_2 emissions between 0.1 and 0.2 per ton (Glavind & Petersen, 2015). Therefore, this result in a huge amount of CO_2 emissions impacting the environment, but the solution for this cannot be an alternative material to concrete; rather it should be a sustainable form of concrete whose impact is lower than conventional concrete. So, Green Concrete can be a solution for this, using residual materials from various other industries and combining them with conventional concrete forms the basic idea of what is Green Concrete. Several materials have suited properties for Green concrete production and have a huge potential to increase material recycling by investigating the possible use these for concrete production. To analyze the environmental impact of green concrete 6 phases need to be considered: '(1) mechanical properties (strength, shrinkage, creep, static behavior, etc.). (2) fire resistance (spalling, heat transfer, etc.). (3) Workmanship (workability, strength development, curing, etc.) (4) durability (corrosion protection, frost, new deterioration mechanisms, etc.). (5) thermodynamic properties (input to the other properties). (6) Environmental aspects (CO_2 emission, energy, recycling, etc.)'. The basic standards that a Green Concrete structure must comply with are: CO_2 emissions shall be reduced by at least 30%; At least 20% of the concrete shall be residual products used as aggregate; Use of concrete industry's residual products; Use of new types of residual products, previously landfilled or disposed of in other ways; CO_2 neutral, waste-derived fuels shall substitute fossil fuels in the cement production by at least 10% (Glavind & Petersen, 2015). Further, Green Concrete is divided into two categories: Aggressive Environmental class (outdoor) and Passive Environmental class (indoor), they both vary in the strength and the days under which the desired strength is achieved.

The reduce, reuse, and recycling techniques, or any two processes in concrete technology, should be followed while making green concrete. The three main goals of the green concrete idea are to minimize greenhouse gas emissions, reduce the use of natural resources such as limestone, shale, clay, natural river sand, and natural rocks that are consumed for human growth but not returned to the earth, and reduce the use of waste materials in concrete that pollute the air, land, and water. This goal of green concrete will lead to long-term development without depletion of natural resources. Some portions of cement can be replaced by fly ash, sludge ash, or any other material with cementitious qualities to obtain green concrete. Fine aggregates can be substituted with quarry dust or iron slag in the needed proportions, while coarse aggregates can be replaced with silica fume, discarded glass, and so on (Goyal & Kumar, 2018).

2. Composition of Green Concrete

Green concrete utilizes at least one of the waste materials such as demolition waste as is a core component for the alternative of cement and Aggregate (Khazaleh & Gopalan, 2019). Here are some alternatives for cement that can be used in the mix design of green concrete:

2.1 Blast Furnace Slag

Green concrete is made from granulated blast furnace slag, steel slag, and fuel gas desulfurization gypsum. The lack of portlandite in this concrete prevents barnacles from fouling it. Algae development is feasible because of the concrete's attachment

2.2   Fly Ash

The replacement of cement with fly ash helps to attain high compressive strength and fracture toughness with particular percentages of 0%, 20%, and 30% as shown by Grzegorz Ludwik Golewski in his tests in 2018.

2.3   Silica Fume

A mixture of silica fumes and marble waste is a potential replacement for cement and they help in improving the strength and durability of the concrete. This also helps to reduce harmful environmental effects by reducing cement consumption by up to 30%.

2.4   Recycled Glass

2.5   Date Palm Ash

The Potential replacements for cement and aggregates are given in Table 1. Followings are some alternatives for aggregate that can be used in the mix design of green concrete:

a. Foundry Sand

Treated used foundry sand can be a good replacement for fine aggregate.

b. Plastic and Demolition Waste

c. Farming Waste

Farming waste such as bamboo, corn wheat, olive, sisal and seashell are great replacements for aggregate.

d. Electronic Waste

Table 1. Potential replacements for cement and aggregates

S. No.	Traditional Ingredients	Replacement Materials for Green Concrete
1	Cement	Eco-Cement, Sludge ash, Municipal solid waste fly ash
2	Coarse Aggregates	Recycled aggregates, waste ready mix concrete, waste glass, recycled aggregates with crushed glass, recycled aggregates with silica fume.
3	Fine Aggregates	Fine recycled aggregates, demolished brick waste, quarry dust, waste glass powder, marble sludge powder, rock dust and pebbles, artificial sand, waste glass, fly ash and micro silica, bottom ash of Municipal solid waste

To understand the mix design of Green concrete certain tests were conducted in which quarry sand and fly ash were the core materials that were used as a replacement for river sand and cement. The two mixes of M40 grade used were '1: 1.39:1.7 (cement: river sand: crushed stone aggregate) and 1: 1.37: 1.7 (cement: quarry dust: crushed stone aggregate)'. The water-cement ratio used in the test for both the mixes was 0.40. To prepare structural grade concrete high volume of high calcium fly ash (ASTM Class C) was used with proportions replacing cement in 40 %, 50%, 60% and 70%. The ratio of fly ash to cement was maintained at 1-1.25 for all tests (Kumar & Moorthy, 2013). Before analyzing the test results here are some preliminary data about the individual properties of the material (Table 2) (Table 3) (Table 4) (Table 5):

Table 2. Physical properties of cement and flyash

Property	Cement	Flyash
Normal consistency	29%	40%
Initial setting time	63minutes	150 minutes
Final setting time	240 minutes	-
Specific gravity	3.15	2.412
The fineness of cement by sieve	1.2%	2.26%

Table 3. Chemical properties of flyash

Properties	Percentage
CaO	12.90
SiO_2	44.50
Al_2O_3	21.10
SO_3	7.81
Na_2O	6.25
K_2O	0.80

Table 4. Physical properties of fine aggregate and coarse aggregate

Property	Fine Aggregate (Sand)	Fine Aggregate (QD)	Fine Aggregate
Specific gravity	2.63	2.41	2.707
Fineness Modulus	2.46	3.77	5.914
Uniformity Coefficient	3.33	9.28	1.479
Coefficient of Curvature	0.948	1.06	1.201

Table 5. Details of mix proportions of concrete

Materials Used	CC	QCC	QCFA₁	QCFA₂	QCFA₃
Fly ash (F.A)%	0	0	10	15	20
Superplasticizer (S.P)%	0	2.0	2.0	2.0	2.0
Cement (Kg/m^3)	530	530	477	450	424
Fly ash (Kg/m^3)	0	0	53	80	106
Sand (Kg/m^3)	740.25	0	0	0	0
Quarry Dust (Kg/m^3)	0	725	725	725	725
Coarse Aggregate(Kg/m^3)	901.53	901.53	901.53	901.53	901.53
Water (lit/m^3)	212	212	212	212	212
Superplasticizer (lit/m^3)	0	10.6	10.6	10.6	10.6

CC=Control concrete without fly ash   QCC=Quarry dust concrete without fly ash
QCFA₁= Quarry dust concrete with 10% flyash, QCFA₂= Quarry dust concrete with 15% flyash QCFA₃= Quarry dust concrete with flyash 20%

The experiment done was by using casting 60 cube and 60 cylinder concrete blocks. The strength, workability, and tensile strength tend to be carried out throughout 3, 7, 14, 28, 60 and 90 days.

Table 6. Results of Workability Tests on Concrete

Type of Concrete	Slump value Mm	Compaction Factor	Flow %
CC	85	0.88	42
QCC	94	0.90	35
QCFA₁	80	0.8	41
QCFA₂	78	0.77	44
QCFA₃	76	0.74	51

We observe the slump value due to the addition of quarry dust decreases rapidly at first and then takes a gradual downfall (Table 6). This is because the water absorption capacity of quarry dust is higher, and further decrease is caused by the properties of fine aggregate (Figure 1).

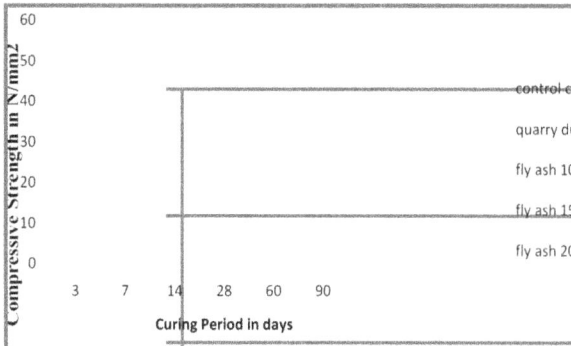

Figure 1. Comparison of compressive strength of cube

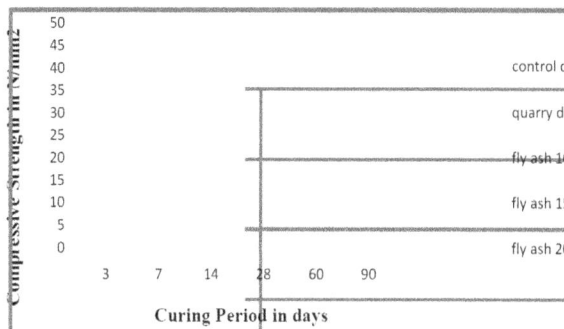

Figure 2. Comparison of compressive strength of cylinder

The compressive strength of QDC is slightly lower than conventional concrete, which is due to the poor particle grading of quarry dust. At 28 days also the cylinder specimen tested for compressive strength showed the same observations. The tensile strength of sand concrete and QDC with fly ash 10% are more or less the same (Figure 2). Therefore, replacing quarry sand with river sand sometimes gives better results in terms of compressive strength. The decrease in early strength of Fly ash concrete can be well compensated by adding quarry dust, while the decrease in workability of QDC can be reduced by adding some amount of fly ash (Figure 3). Hence we can replace conventional river sand to quarry dust with a partial fly ash mixture, which can be observed in the experiment (Kumar & Moorthi, 2013). The initial setting time of green concrete specifically flyash-based concrete increases from 150 min to 200-250 min based on the percentage of fly ash being used in the replacement of cement.

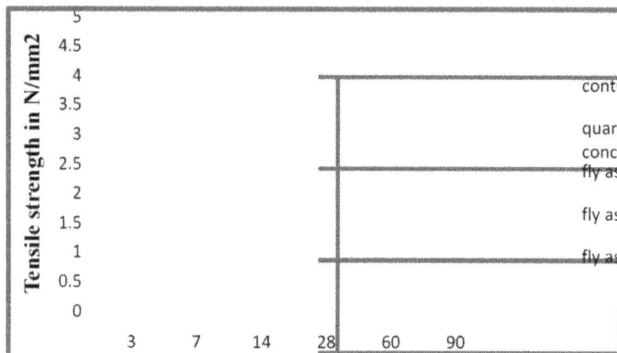

Figure 3. Comparison of split tensile strength of cylinder

3.   Cost Analysis of Green concrete and Conventional Concrete

Based on the maximum percentage of replacement and compressive strength, the cost analysis of two concrete mixtures with the replacements of cement with fly ash and fine aggregate with bottom ash for the binder content of 333 kg/m3; and fly ash for cement, bottom ash for fine aggregate, and fly ash aggregate for coarse aggregate replacements for the binder content of 389 kg/m3 were evaluated. A cost analysis was performed between normal and green concrete with fly ash bottom ash and fly ash aggregate with the help of relevant studies for evaluating the economic viability of concrete.

3.1   *Cost Analysis of Green Concrete with Fly Ash and Bottom Fly Ash with the Cement of 333 kg/m³*

The cost of 1 m³ of concrete was calculated for both conventional and green concrete. The cost study was done based on the proportion of replacement for cement and fine aggregate, with green concrete compressive strength ranging from 28 to 180 days. The cost of building materials such as cement, aggregates, and HRWRA was taken into account. The compressive strength at 28 to 180 days was utilized for the cost study since green concrete showed strength development as curing progressed. The cost study was based on the amount of replacement and the strength properties of fly ash and bottom ash in concrete. The cost of 30 % fly ash, 30% bottom ash, and 30 % fly ash plus 30 % bottom ash was investigated (Table 7).

Table 7. Cost analysis between normal and green concrete (30 FA, 30 BA and 30FA30BA)(Cement content of 333 kg/m³)

Material	Units	Rate in INR	Normal concrete		Green Concrete					
			30 MPa		30FA [50-60MPa]		30BA [40-55MPa]		30FA30BA [40-60 MPa]	
			Quantity per 1m³ (kg)	Rate (INR)	Quantity per 1m³ (kg)	Rate (INR)	Quantity per 1m³ (kg)	Rate (INR)	Quantity per 1m³ (kg)	Rate (INR)
Cement	Bag	290	333.0	1931.40	233.1	1351.98	333.0	1931.4	233.1	1351.98
Fly ash	m³	65	0.0	0.00	99.9	2.32	0.0	0.0	99.9	2.32
20mm	m³	940	516.3	175.21	516.3	175.21	516.3	175.2	516.3	175.21
12mm	m³	871	774.4	249.84	774.4	249.84	774.4	249.8	774.4	249.84
Sand	m³	110	754.0	31.24	754.0	31.42	527.8	21.9	527.8	21.99
Bottom ash	m³	65	0.0	0.00	0.0	0.00	204.0	5.5	204.0	5.57
Water	liter	0	0.0	0.00	0.0	0.00	0.0	0.0	0.0	0.00
HRWRA	liter	180	1.1	209.79	1.6	300.60	1.6	300.60	1.6	300.60
Total cost				2597.66		2111.37		2684.62		2107.52
% Savings				-		18%		-3.35%		19%

INR=Indian Rupee, FA=Flyash, BA=Bottom ash, HRWRA-High range water reducing admixture

Table 8. Cost analysis between normal and green concrete (40FA, 100BA and 40FA30BA, 40FA60BA, 40FA100BA) (Cement content of 333 kg/m³)

Material	Units	Rate in INR	Normal concrete		Green Concrete							
			30 MPa		40FA [50 -60MPa]		100BA [40-55MPa]		40FA30BA [40-60 MPa]		40FA60BA [30-40MPa]	
			Quantity per 1m³ (kg)	Rate (INR)	Quantity per 1m³ (kg)	Rate (INR)	Quantity per 1m³ (kg)	Rate (INR)	Quantity per 1m³ (kg)	Rate (INR)	Quantity per 1m³ (kg)	Rate (INR)
Cement	Bag	290	333.0	1931.4	199.8	1158.84	333.0	1931.4	199.8	1158.84	199.8	1158.84
Fly ash	m³	65	0.0	0.0	133.2	3.09	0.0	0.0	133.2	3.09	133.2	3.09
20mm	m³	940	516.3	175.21	516.3	175.21	516.3	175.21	516.3	175.21	516.3	175.21
12mm	m³	871	714.4	249.84	714.4	249.84	714.4	249.84	714.4	249.84	714.4	249.84
Sand	m³	110	754.0	31.42	754.0	31.42	0.0	0.00	527.8	21.99	301.6	12.57
Bottom ash	m³	65	0.0	0.00	0.0	0.00	680.0	18.57	204.0	5.57	408.0	11..14
Water	liter	0	0.0	0.00	0.0	0.00	0.0	0.00	0.0	0.0	0.00	0.00
RWRA	liter	180	1.1	209.79	1.8	329.67	5.5	1006.20	1.8	329.67	4.1	748.80
Total cost				2597.66		1948.07		3381.23		1944.22		2359.50
% Savings				-		25%		-7.39%		25%		9%

INR=Indian Rupee, FA=Flyash, BA=Bottom ash, HRWRA-High range water reducing admixture

In the above analysis fly, ash and bottom ash were shown to be cost-effective alternatives for cement and fine aggregate in the cost analysis for green concrete with a binder content of 333 kg/m^3, which may be designed for 25 to 55 MPa (Table 8). As a result of its water absorption capability in concrete, 100% bottom ash raised the dose of HRWRA but removed the usage of natural river sand in construction.

3.2 Cost Analysis of Green Concrete with Fly Ash, Bottom Ash and Fly Ash Aggregate with a Cement Content of 389 kg/m^3

The cost of calcium hydroxide, cement, fly ash, and the electricity used by the pelletizer to make fly ash pellets are all taken into account in the green concrete cost analysis. For 100 kg of fly ash pellets, the cost to make them was 50 INR. Green concrete with 30 percent fly ash, 30 percent bottom ash, and 20 percent fly ash aggregate cost 2633.93 INR, which was 14 percent less than the cost of normal concrete, which cost 3066.66 INR. There was a 14% savings in the cost of building the green concrete that had 30% fly ash, 30% bottom ash, and 20% fly ash aggregate (Table 9). The concrete had a compressive strength of 45 to 50 MPa (28 to 180 days).

Table 9. Cost analysis of the green concrete with fly ash, bottom ash and fly ash aggregate (cement content of 389 kg/m3)

Material	Units	Rate in INR	Normal concrete		Eco-friendly Green Concrete 30 FA 30 BA 20 FAA [45-50 Mpa]	
			Quantity per 1m^3 (kg)	Rate (INR)	Quantity per 1m^3 (kg)	Rate (INR)
Cement	Bag	290	389.0	2256.20	272.3	1579.34
Fly ash	m^3	65	0.0	0.00	116.7	2.71
20mm	m^3	940	511.4	173.54	409.1	138.83
12mm	m^3	871	767.0	247.43	613.6	197.94
FAA 12 mm	m^3	50	0.0	0.00	117.4	58.71
FAA 20 mm	m^3	50	0.0	0.00	78.2	39.14
Sand	m^3	110	715.6	29.82	302.0	12.58
Bottom Ash	m^3	65	0.0	0.00	193.5	5.29
Water	Liter	0	0.0	0.00	0.0	0.00
HRWRA	Liter	180	1.1	359.64	3.3	599.4
Total cost				3066.63		2633.93
% Savings				-		14%

INR-Indian Rupee, FA-Flyash, BA-Bottom ash, FAA-Fly ash aggregate
HRWRA-High range water reducing admixture

4. Conclusions

The cost study of green concrete revealed that fly ash, bottom ash, and fly ash aggregate improved technical, environmental, and economic efficiency without sacrificing strength and durability. This method of building replaces cement with fly ash, fine aggregate with bottom ash, and coarse aggregate with fly ash aggregate. This will provide the groundwork for future research by lowering the use of natural river sand, reducing the quarrying of natural rocks, and protecting natural land for the disposal of fly ash and bottom ash. Using fly ash, bottom ash, and fly ash aggregate in buildings encourages green technology and sustainable development for future generations. Therefore, concluding that green concrete with a cement weight of 333 kg/m3 saved up to 25% of the cost of conventional concrete in 6.3.3.1, but only 14% in 6.3.3.2 with a cement content of 389 kg/m3. With the addition of HRWRA and an extended curing time, green concrete made of fly ash, bottom ash, and fly ash aggregate may be constructed to be cost-efficient and environmentally beneficial without sacrificing the strength and durability of conventional concrete.

CHAPTER 6

Self-Healing Concrete

1.   Introduction

Concrete is the most often utilized building material. Concrete is brittle in tension yet robust in compression, thus fractures are unavoidable. When fractures appear in concrete, the lifetime of the concrete may be shortened. Microcracks and pores in concrete are very undesirable because they create an open channel for water and other harmful elements to enter, resulting in reinforcement corrosion and a reduction in the strength and durability of the concrete. The fissures may be repaired using a variety of procedures, but they are both costly and time-consuming. Self-Healing Concrete is a reasonable technology for repairing fractures in concrete by itself. This bacterial remediation technology outperforms others since it is bio-based, environmentally benign, cost-effective, and long-lasting. Because concrete is an alkaline substance, bacteria introduced to it must be able to resist the alkali environment and most of the microorganisms die at a pH value of 10. Just at the time of mixing, bacteria with a calcium nutrition source are introduced to the concrete. If there are any fractures in the concrete, bacteria will precipitate calcium carbonate. The fractures will be sealed as a result of this. The synthesis of urease enzyme by urease-positive bacteria has been demonstrated to impact the precipitation of calcium carbonate (calcite). Calcite precipitation's pH rises as a consequence of this. Fractures larger than 0.8mm are more difficult to cure, yet cracks may heal with calcite precipitation when bacteria are used. Some of the "Urease positive bacterium are – Bacillus megaterium, Bacillus pasteurii, Bacillus sp. CT-5, Bacillus subtilis, Bacillus aerius, Sporosarcina pasteurii, AKKR5, Shewanella Species, Bacillus flexus etc". The bacteria-based self-healing agent is supposed to be able to hibernate for up to 200 years beneath the concrete. When bacterial spores come into touch with water and oxygen as a result of concrete fractures, they begin a microbial activity. Lately, self-healing techniques have shown favorable results in repairing fractures that are still in the initial stages of development. Also, the cracks must not exceed a depth of 150mm to get outstanding results (K., et al., 2018).

2.   Process Involved in Self-Healing Concrete

There are three processes of self-healing concrete which comprise of (i) Natural Process, (ii) Chemical process, and (iii) Biological process.

2.1   The Natural Process

This process can partially fix the cracks in concrete, and it is divided into the following four techniques (Figure 1):

a) Formation of $CaCO_3$ or CaOH (calcium carbonate or calcium hydroxide)

b) Crack is obstructed by impurities in the carriage of water

c) Crack is further obstructed by hydration of the unreacted cement

d) Crack is impeded by the enlargement of hydrated cementitious pattern in the crack loins (such as the lump of calcium silicate hydrate gel)

Multiple events may include the occurrence of more than one of these processes simultaneously. Perhaps the majority of these systems can only partially close up the openings of certain fractures and cannot completely close them off. This will aid in halting the growth of fractures and preventing the inward entrance of corrosive compounds such as acids. Among the self-healing methods advocated in natural processes, calcium carbonate and calcium hydroxide configurations are the most common and effective strategies for spontaneously repairing concrete (Magaji, et al., 2019).

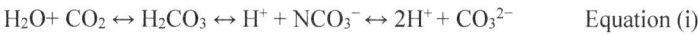

$$H_2O + CO_2 \leftrightarrow H_2CO_3 \leftrightarrow H^+ + NCO_3^- \leftrightarrow 2H^+ + CO_3^{2-} \qquad \text{Equation (i)}$$

Loose calcium ions are emitted into concrete as a result of cement hydration and dissipation and are countered by NCO_3^- and CO_3^{2-} at cracking surfaces. As a result, calcium carbonate crystals form. Reactions (ii) and (iii) may occur only at pH values greater than or equal to 7.5. Crystals form along the surface of the fractures and eventually permeate the space (Magaji, et al., 2019).

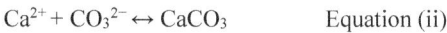

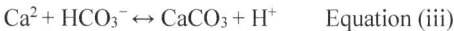

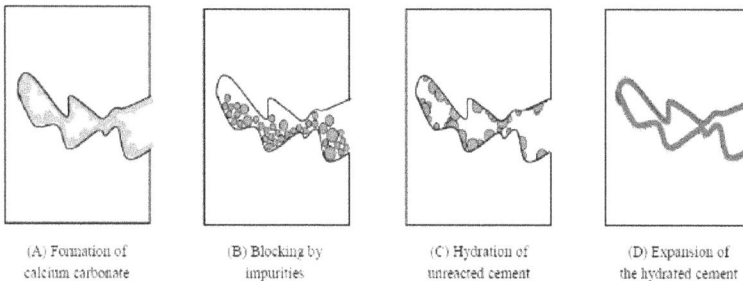

$$Ca^{2+} + CO_3^{2-} \leftrightarrow CaCO_3 \qquad \text{Equation (ii)}$$

$$Ca^2 + HCO_3^- \leftrightarrow CaCO_3 + H^+ \qquad \text{Equation (iii)}$$

(A) Formation of calcium carbonate

(B) Blocking by impurities

(C) Hydration of unreacted cement

(D) Expansion of the hydrated cement

Figure 1. Types of natural processes in self-healing concrete

2.2 The Chemical Self-Healing Process

The chemical healing is a term that refers to artificial healing which uses chemical composites. The chemical liquid reagents (glue) are mixed with fresh concrete in small containers to create self-healing concrete. There are two types of chemical processes: a) Hollow pipettes and vessel networks containing glue (Figure 2); b) Encapsulated Glue (Figure 3).

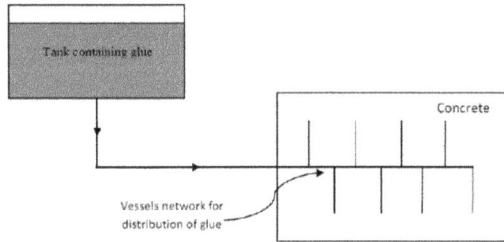

Figure 2. Hollow pipettes and vessel networks containing glue

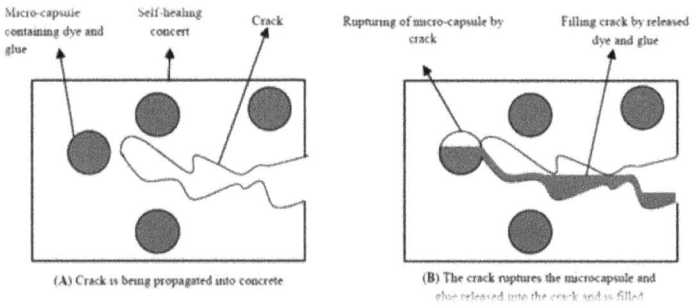

Figure 3. Encapsulated microcapsules glue

2.3 The Biological Self-Healing Process

A lot of places can have microorganisms. They can be found in water, soil, oil, acidic hot springs, and industrial waste. Microorganisms are usually broken down into three main groups: bacteria, fungi, and viruses. These microorganisms are used to make the genetic self-healing concrete. Some of these bacteria can make certain chemicals, so they are used to make the concrete. Different ways may be used to include microorganisms into the biological self-healing concrete. These include the direct application of microbial brew to new concrete arrangements, as described under the chemical procedure for sharing microorganisms. The pH, temperature, and moisture content of concrete are usually not conducive to bacterial growth. As a result, rather than using fresh microbial broth, the resistant kind of bacteria (spore) is used in certain circumstances. Encapsulated microbes, on the other hand, may be used to withstand the severe conditions of concrete. There are two biological processes under this category: a) Precipitation of calcium carbonate; b) Precipitation of polymorphic iron aluminum silicate. These methods are usually conducted by either fungi or bacteria and apart from these, two different types of microorganisms play a significant role in self-healing processes i.e. Mesophilic microorganisms and Thermophilic microorganisms, which are further divided into Aerobic and Anaerobic microorganisms (Magaji, et al., 2019).

3. Classification of Self-Healing Concrete

3.1 Autogenous Self-Healing Concrete

The majority of autogenous self-healing is dependent on increased hydration of concrete, calcium hydroxide carbonation, and another binder.

(a) Blocking cracks by waste

(b) Carbonation of CaOH

(c) Expansion of the hydrated concrete matrix in crack flanks

(d) Ongoing hydration of clinker minerals cracks

3.2 Autonomous Self-Healing Concrete

Autonomous self-healing concrete was fully dependent on a physical process. The phrase 'autonomous self-healing' has been used to refer to this phenomenon.

(a) The vascular method;

(b) The capsule method;

(c) The bacterial method;

(d) The electro-deposition method;

(e) The shape memory alloy method;

(f) The microwave method and/or induction energy

4.   The Environmental Impact of Self-Healing Concrete

Self-healing concrete minimizes the quantity of carbon dioxide released into the atmosphere as a consequence of concrete manufacture. This is because concrete manufacture is energy demanding in several ways, especially when transportation, mining, and concrete plants are taken into account. However, industries are the primary sources of carbon dioxide emissions in India, accounting for around 8% of total emissions. In terms of increasing the lifetime of concrete as well as reducing maintenance, self-healing concrete will minimize the creation of surplus concrete, hence lowering carbon dioxide emissions in our environment (Magaji, et al., 2019).

5.   Composition / Mix Design of Self-Healing Concrete

To understand the composition and mix design of self-healing concrete, an experiment was conducted based upon a comparative study between conventional concrete and bacterial concrete of the same grade (Table 1). The materials used were:

(a) Ordinary Portland cement of grade 53 as per IS: 12269 (1987b)

(b) River sand was chosen, which passed through a 4.75mm IS sieve and confirmed to zone-1 of IS: 383 (1987a). It was observed that the specific gravity was 2.3.

(c) Coarse aggregate: Crushed stones up to 20mm in size are retained on 4.75mm IS sieves. 3.13 were discovered to be the specific gravity.

(d) Potable water for conventional concrete

(e) Bacterial water - consisting of 10^5 cells of Bacillus megaterium / ml of water

(f) A thin metal sheet with a thickness of 0.3mm is used to create an artificial fracture up to a depth of 10mm in an unhardened concrete specimen.

Table 1. Mix design of M25 concrete using IS:10262 (2009) and IS:456 (2000)

Ingredients	Cement	Fine aggregate	Coarse aggregate	Water
Quantity (Kg/m3)	340	657.6	1335.94	171.7
Ratio	1	1.93	3.93	0.51

The design mix for conventional and bacterial concrete used was M25 grade with the ratio of 1:1.9:3.9 (cement: Fine aggregate: Coarse aggregate) having a water ratio of 0.51. For the bacterial concrete, the bacteria water replaces the potable water with an M25 concrete grade. Now there is a need to identify a suitable bacterium that would sustain the alkali environment of concrete, hence there are certain bacteria that can be considered:

- Bacillus megaterium
- Bacillus pasteurii
- Bacillus sp. CT-5
- Bacillus subtilis
- Bacillus aerius
- Sporosarcina pasteurii
- AKKR5
- Shewanella Species
- Bacillus flexus

When compared to other urease-positive bacteria, Bacillus megaterium may precipitate the most quantity of calcite, resulting in a greater improvement in compressive strength and crack-healing efficiency.

6. Mechanism of Self-Healing Concrete

Bacterial water contains 10^5 Bacillus megaterium cells per milliliter of water. The bacterium enters a latent condition, and when fractures appear in the future, the bacteria are exposed to air and water, and they begin to precipitate calcite crystals. Such bacteria's spores have thick cell that allows surviving up to 200 years while waiting for a favorable environment to germinate (Figure 4). Calcite precipitation is affected by bacteria decomposing urea via the bacterial urease enzyme. The metabolism of bacteria produces urease, which catalyzes the conversion of urea to ammonia and carbonate. These components also hydrolyze into carbonic acid and ammonium chloride, resulting in calcium carbonate (Figure 5). The surface of bacteria, which is negatively charged and has a neutral pH, plays a crucial role in calcite precipitation. The positive-charged calcium ion may mix with bacteria's surface, promoting nucleation (K., et al., 2018).

$$CO(NH_2)_2 + H_2O \rightarrow NH_2COOH + NH_3 \qquad (1)$$

$$NH_2COOH + H_2O \rightarrow NH_3 + H_2CO_3 \qquad (2)$$

$$H_2CO_3 \leftrightarrow HCO_3^- + H^+ \qquad (3)$$

$$2NH_3 + 2H_2O \leftrightarrow 2NH_4^+ + 2OH^- \qquad (4)$$

$$HCO_3^- + H^+ + 2NH_4^+ + 2OH^- \leftrightarrow CO_3^{2-} + 2NH_4^+ + 2H_2O \quad (5)$$

$$Ca^{2+} + Cell \rightarrow cell - Ca^{2+} \qquad (6)$$

$$Cell - Ca^{2+} + CO_3^{2-} \rightarrow Cell - CaCO_3 \downarrow \qquad (7)$$

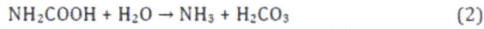

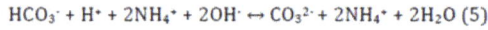

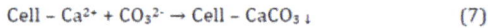

Figure 4. Chemical equation depicting the formation of calcite crystal

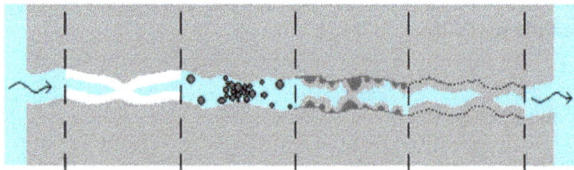

Figure 5. Healing mechanism of bacterial concrete

7. Properties and Experimental Results

7.1 Compressive Strength

When comparing compressive strength, it was evident that bacterial concrete outperformed normal concrete (Figure 6). The compressive strength of bacterial concrete was found to be 11.96 percent higher than that of conventional concrete.

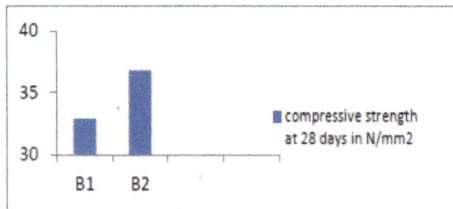

Figure 6. Comparison of compressive strength between conventional concrete [B1] and bacterial concrete [B2]

7.2   *Water Absorption*

Because of the precipitation of calcite on the surface of the specimen, the water absorption of the bacterial concrete surface has improved when compared to conventional concrete (Figure 7). Water absorption was reported to have decreased by 0.45%.

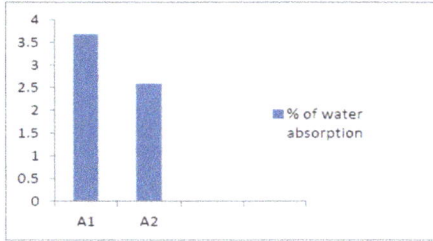

Figure 7. Comparison of water absorption between conventional concrete [A1] and bacterial concrete [A2]

7.3   *Water Permeability*

Due to the plugging of micro-pores by calcite, the depth of water penetration in bacterial concrete is also reduced as compared to ordinary concrete (Figure 8).

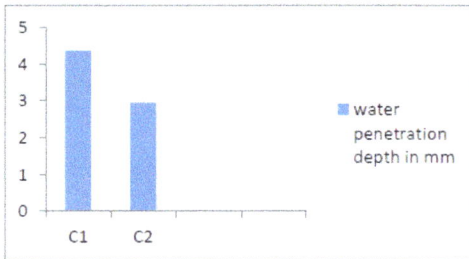

Figure 8. Comparison of water penetration between conventional concrete [C1] and bacterial concrete [C2].

7.4   *Other Experimentation Results*

By incorporating bacteria into the concrete, the compressive strength of the concrete is increased in comparison to normal concrete (Table 2). By adding bacillus subtilisjc3 to ordinary concrete, the compressive strength of the concrete was increased by 14.92 %. In comparison to ordinary concrete, B-Sphaericus increased the compressive strength of concrete by 30.76 % after three days, 46.15 % after seven days, and 32.21 % after 28 days (Luhar & Gourav, 2015).

Table 2. Comparison of compressive strength of conventional concrete and bacterial concrete

S. No.	No. of days	Compressive strength of conventional concrete cubes, N/mm^2	Compressive strength of B- Sphaericus concrete cubes, N/mm^2	% increase in strength
1	3	19.24	25.16	30.76
2	7	23.66	34.58	46.15
3	28	34.52	45.72	32.21

Tensile strength refers to a material's capacity to resist a pulling (tensile) force. Earlier research has shown that bacterial concrete has a higher tensile strength than ordinary concrete as shown in Table 3. The Table 4, Table 5, Table 6 and Table 7 shows the properties, experimentation and results of self-healing concrete.

Table 3. Comparison of tensile strength of conventional concrete and bacterial concrete

S. No.	No. of days	Compressive strength of conventional concrete cubes, N/mm^2	Compressive strength of B- Sphaericus concrete cubes, N/mm^2	% increase in strength
1	3	3.78	4.30	13.75
2	7	4.62	5.28	14.28
3	28	4.85	5.74	18.35

Table 4. Bacteria other than Bacillus can survive in the alkaline environment

S. No.	Application	Types of Bacteria
1.	As a crack healer	B. pasteurtii
		Deleya Halophila
		Halomonasrurihalina
		Myxococcus Xanthus
		B. memgaterium
2.	For surface treatment	B. sphaericus
3.	B. spharicus	Bacilllussubitilis
		B. spharicus
		Thiobacillus

Table 5. Various types of bacteria and their compressive strength results

Bacteria used	Best Results	Bacterial concentration
Bacillus sp. CT-5	Compressive strength is 40% more than the control concrete	5×10^7 cells/mm^3
Bacillus megaterium	Max. The rate of strength development was 24% achieved in the highest grade of concrete 50 Mpa	30×10^6 cfu/ml
Bacillus subtilis	Improvement of 12% in compressive strength as compared to controlled specimens with lightweight aggregates	2.8×10^8 cells/ml
Bacillus aerius	Increase in compressive strength by 11.8% in bacterial concrete compared to control with a 10% dosage of RHA	10^5 cells/ml
Sporosarcina pasteurii	Compressive strength 355 more than the control concrete	10^5 cells/ml
AKKRS	10% increase in compressive strength as compared to control concrete	10^5 cells/ml
Shewnella Species	25% increase in compressive strength of cement mortar compared with the control mortar	100,000 cells/ml

Table 6. Self-healing techniques and measured variable

Approach	Crack depth and width
Micro-encapsulation	The maximum depth of 35 mm crack was filled
Bacteria direct application	A maximum depth of 27.2 mm was filled
Bacteria and Encapsulation	Healing of maximum crack width of 0.970 mm was reported

Table 7. Summarized contrast between specific techniques

Strategy	Advantage	Disadvantage
Bacteria	Biological activities and pollution-free and natural way	The measure should be taken to protect the bacteria in concrete. Many prerequisites to be meet
Encapsulation	Healing agent discharge on the requirement	Complexity in casting
	Potential effectiveness under many damage measure	The possible difficulty of healing agent release

8. Factors Affecting Self-Healing Concrete

The five main factors that affect the self-healing concrete are:

8.1 Moisture Content

The pilot specimen is maintained in water and heals more quickly on its own.

8.2 Crack Width

The cracks with a width of less than 0.3 mm may be healed. The cracks larger than 0.3 mm may not heal. After roughly 200 hours, cracks with a width of 0.1 mm are entirely healed. Furthermore, cracks with a width of 0.2 and 0.3 mm heal in around 30 days. Cracks with a width of 0.15 to 0.3 mm shrink considerably in 7 days and heal entirely in 33 days.

8.3 Time for Hydration

The hydration over a longer period may result in improved self-healing.

8.4 Pressure Applied to Cracks

Applying the right amount of pressure to cracks may help them heal faster.

8.5 Water-Cement Ratio

A greater water-cement ratio means more unreacted cement particles may be utilized for subsequent hydration to improve calcium carbonate production. Moreover, the fracturing interval is significant. Because early breaking concrete has more unreacted cement particles, it has a higher self-healing capacity while hydration is maintained. Also cracks up to 1mm wide can be filled using adhesive-filled fibers and bacteria or fungi which remain dormant for hundreds of years within the concrete and get activated when coming into contact with water and other gases (Huang & Kaewunruen, 2020).

9. Advantages of Self-Healing Concrete

The following are the advantages of using self-healing concrete:

- Concrete's strength is considerably increased when self-healing concrete is used.
- When compared to normal concrete, it has a lesser permeability.
- Additionally, it has a lower water absorption rate than traditional concrete.
- It is very resistant to freeze and thaw assaults.
- Corrosion of reinforcement is almost eliminated
- Repairing cracks may be accomplished effectively.
- The overall cost of upkeep for this concrete is minimal.

10. Disadvantages of Self-Healing Concrete

The following are the disadvantages of using self-healing concrete:

- IS codes and other codes do not refer to the design of microbiological concrete.
- The cost of this concrete is comparable to regular concrete; it is around 10%-30% more expensive.
- Bacterial germination does not occur in all bacteria.
- Costly research is required to detect calcite precipitation.
- Bacteria employed in concrete are harmful to human health; thus, their use should be confined to construction.

11. Cost Analysis of Self-Healing Concrete and Conventional Concrete

The cost analysis for self-healing and conventional concrete is conducted. Where it is observed that the cost of conventional concrete with a binder content of 333 kg/m^3 is 2598 rupee per cubic meter. While according to Dr. Henk Jonkers the cost of producing self-healing concrete is quite high i.e., 13710 rupees per cubic meter, this would be a feasible product only in specific civil engineering constructions where the cost of concrete is much higher because of its superior quality, such as tunnel linings and maritime structures where safety is a major consideration – or in structures with restricted access for repair and maintenance. In these instances, the cost increase associated with the introduction of self-healing agents should be manageable. It is believed that if self-healing concrete is created on a large scale, the cost might be significantly reduced. If the structure's life can be prolonged by 30%, the doubling of the expense of actual concrete will still save a significant amount of money in the long run (Arnold, 2011).

A second self-healing chemical is being developed, which will be significantly cheaper and result in much stronger concrete. The calcium lactate, which is now highly costly, accounts for the bulk of the additional expense. Because it includes a vacuum technology, the process of embedding the bacteria and nutrients into the pellets is also costly. The cost of self-healing concrete might be reduced to Rs. 7283-7712 per cubic meter if a sugar-based dietary component is used (Table 8).

Table 8. Cost analysis of conventional concrete

Material	Unit per	Rate in INR	Normal concrete	
			30 Mpa	
			Quantity Per 1m³ (kg)	Rate in INR
Cement	Bag	290	333.0	1931.40
20 mm	m³	940	516.3	175.21
12 mm	m³	871	774.4	249.84
Sand	m³	110	754.0	31.42
Water	liter	0	0.0	0.00
HRWRA	liter	180	1.1	209.79
Total cost				2597.66
% Savings				-

12. Conclusions

From the above analysis, it is clear that we as a generation need a greener construction material that can act as a replacement for conventional concrete. In both cases, we have seen that property-wise, whether it be compressive strength, workability, or tensile strength, conventional concrete is far out casted by Green and Self-healing concrete. Talking about the cost, then green concrete is the ultimate winner but self-healing concrete although costlier than conventional concrete still has more pros than cons. In the long run mainly for the big infrastructure development projects, it can be a game-changer though studies are still underway to convert this concrete to a larger contributor in the construction industry. Both concrete has the potential to reduce the harmful environmental impact because the use of residual waste helps in the reduction of embodied energy of concrete and prevents landfills. They reduce cement production to a great extent by increasing the lifespan of the building. Hence, these materials must be used more often on the construction sites to be more environmental friendly.

CHAPTER 7

Conclusions

1. Introduction

3D concrete printing is one of the most efficient forms of construction technology. It can be very well described as a clean technology as compared to today's conventional techniques. Indeed, this technique has not yet hit the construction market but sooner or later it is going to revolutionize the industry to a large extent. Today precast and pre-stress are the technologies that have taken over the stakeholders but as the upcoming generation, we as human beings are more aware of the environment, we care for it and henceforth we have to work for its betterment.

The initial cost of concrete printing technology is high due to the high cost of 3D printer machines. 3D concrete printing provides many potential cost-effectiveness advantages for construction processes compared to conventional construction methods, taking into account the different cost elements (labor, equipment, materials, design and planning costs). Design optimization in 3D technology increases the complexity of the shape but also reduces the use of material when considering the impact on the environment. As a consequence, 3D engineering is expected to achieve better environmental quality over the entire service life of structures with the same features.

2. Combining 3D-Printing Technology with Green and Self-Healing Concrete

3D printing already being efficient when configured or combined with sustainable forms of material i.e. Green and self-healing concrete has a great potential to be one of the most eco-friendly forms of construction. Some projected advantages of this fusion are:

(a) Decreased carbon footprint by using recycled waste material in green concrete

(b) Enhanced life cycle of the structure due to self-healing concrete which eventually cuts down the production of cement, aggregate reduces the CO_2 content in the atmosphere

(c) Low material wastage in 3D Concrete Printing, which means less material consumed, less material produced, and the ones which are produced use waste material, creating an interdependent cycle of a process that helps in reducing the CO_2 emissions.

(d) Faster construction and small-scale affordable projects can help meet the Global sustainable goal of housing for all.

(e) Better structural performance as compared to conventional technique and a great cost savior

(f) Increased job opportunities for skilled labor.

(g) Reduced on-site injuries

(h) Helps in avoiding landfills by using waste material in construction and eventually saves our flora and fauna

(i) Helps in achieving peculiar design form i.e. gives design freedom

3.   Opportunities and Future Implications

The 3D printing gives the construction industry, which is always considered to be an industry with minimal mass customization, a lot of mass customization alternatives. The construction sector can be opened up to a wide range of product customization contrary to conventional methods of restricting the creativity of architects. The effectiveness of 3D printing in the construction industry depends on two factors: the nature of the needs of the consumer and the degree of the desire of the customer. For example, mass customization in the house building sector has been the key marketing strategy in the construction industry in Korea for the past 30 years. Researchers agree that applications for construction using 3D printing technology will gradually increase shortly (Wu et al., 2016). In addition, 3D printing technology can easily produce a complex geometry structure that is difficult and expensive to build using conventional methods using standard construction procedures. 3D printing technology will accelerate the production of revolutionary constructions that are not technically and economically available a few years ago. New structural models need to be explored, which will increase the potential of 3D printing technology in the construction sector and demonstrate their benefits (Camacho et al., 2018). This brings many benefits, such as reducing material consumption and the impact on the environment, achieving esthetic appearance and limitlessness Designs. It must be kept in mind that 3D-based buildings must comply with structural rules to maintain the levels of safety and quality required by current building codes. The method will encourage more innovative designs, but they will have to be more rational (Labonnote et al., 2016).

Back in 2004 when the 3D Concrete Printing technology first emerged, it was only possible to build a wall but today the same technology can build a house in just 24 hours. The world recently faced the lockdown due to a pandemic and it was just the beginning. Nature has just started to show its color, it has started to return everything humans have given and now it's time to realize that Global warming and climate change are slow poison. India the second-largest producer of cement worldwide needs to anyhow reduce production and adopt more sustainable forms of construction. The study suggests that 3D concrete printing is a very clean and sustainable method of building. It has strategically proven how green and self-healing concrete can be the next step of advancement; it would help in reducing the CO_2 content largely. Although Precast, pre-stress and other conventional technologies right now cannot be eradicated because 3D printing has not yet proven effective in any study for large-span structures or transportation structures but it is a great replacement for the residential sector. Further 3D printing is right now being tested for high-rise modular structures but studies are yet to come, till then precast and other technologies are the major choices. The cost of self-healing concrete is currently a challenge, due to the high cost of calcium carbonate, its use practically in the industry is not yet feasible, though studies by many scientists are underway to make it the next conventional form of concrete. Green concrete is one such material that currently has great potential whether with or without the 3D Concrete Printing, it is self-sustainable and has proven to be one step ahead of the common concrete. So, this is something that can be implemented on a larger scale to reduce the carbon footprint of India by 2030.

Much more research is needed on still-unsolved issues such as structural and mechanical stability, material life, and toxic effects of materials, to name a few. Continuous research will focus on interdisciplinary work involving the science of materials, methods of production, robotics, architecture and design to solve these problems. Once problems with 3D concrete printing technology are overcome, 3D printing in the field of construction will achieve its full prospect. The life cycle performance of printed buildings/building components is currently unknown, especially because 3D printing in the construction industry is still in its infancy. It is plausible to argue that by focusing on these issues, 3D printing will soon attain its full potential in the building business.

Bibliography

[1] Agarwal, N. and Garg, N. (2018) A Research on Green Concrete. IJIRMPS, 6(4), pp. 362-378.

[2] Albrecht, L (2021) Power blanket. [Online] Information on https://www.powerblanket.com/blog/long-concrete-take-set/

[3] Allouzi R., Al-Azhari W. and Allouzi R. (2020) Conventional Construction and 3D Printing: A Comparison Study on Material Cost in Jordan. Journal of Engineering, 2020(Article ID 1424682), pp. 1-14. https://doi.org/10.1155/2020/1424682

[4] Anon (2021) Designing buildings wiki. [Online] Information on https://www.designingbuildings.co.uk/wiki/Prestressed_concrete

[5] Arnold D., 2011. Self-Healing Concrete. Ingenia, 1(46), pp. 39-43.

[6] Aslam Hossain M., Zhumabekova A., Chandra Paul S. and Ryeol Kim J. (2020) A Review of 3D Printing in Construction and its Impact on the Labor Market, pp. 1-21. https://doi.org/10.3390/su12208492

[7] Baikerikar A. (2014) A Review on Green Concrete. Journal of Emerging Technologies and Innovative Research, 1(6), pp. 472-474.

[8] Betsky A. (2014) Architect magazine. [Online] Information on https://www.architectmagazine.com/technology/the-future-of-3d-printing-in-the-construction-industry_o

[9] BMTPC (2019) Building Materials and Technology Promotion Council, [Online] Information on https://bmtpc.org/DataFiles/CMS/file/PDF_Files/61_PAC_Urbaanic_Final.pdf

[10] Bos F., Wolfs R., Ahmed Z. and Salet T. (2016) Additive manufacturing of concrete in construction: potentials and challenges of 3D concrete printing. Virtual and Physical Prototyping, 11(3), pp. 209-225. https://doi.org/10.1080/17452759.2016.1209867

[11] Camacho D. D., Clayton P., O'Brien W. J., Seepersad C., Juenger M., Ferron R. and Salamone S. (2018) Applications of additive manufacturing in the construction industry A forward-looking review. Automation in construction, 89, pp. 110-119. https://doi.org/10.1016/j.autcon.2017.12.031

[12] Chen D. (2015) Direct digital manufacturing: definition, evolution, and sustainability implications, Journal of Cleaner Production, Elsevier, 107, pp. 615-625. https://doi.org/10.1016/j.jclepro.2015.05.009

[13] Chudaman P. P. H., Dagdu P. W. H. and Basharatkha P. S. (2015) Evolution of Construction Technique: A Literature Review. International Journal of Latest Trends in Engineering and Technology, 5(3), pp. 52-56.

[14] Civilsir (2021) Civil Sir. [Online] Information on https://civilsir.com/rate-analysis-for-m20-concrete/

[15] CPWD (2018) Schedule of Rates 2018- For New and Innovative Technologies, Central Public Works Department, New Delhi, India.

[16] CPWD (2021) Analysis of Rates for Delhi (Vol-1), New Delhi: Authority of Director General, CPWD.

[14] El Sayegh S., Romdhane L. and Manjikian S. (2020) A critical review of 3D printing in construction: benefits, challenges, and risks. Archives of Civil and Mechanical Engineering , pp. 1-25. https://doi.org/10.1007/s43452-020-00038-w

[15] Glavind M. and Petersen C. M. (2015) Green Concrete-A Life Cycle Approach In: Glavind M. and Petersen C. M., eds. Challenges of Concrete Construction: Volume 5, Sustainable Concrete Construction. Denmark: ICE, pp. 19-25.

[16] Goyal M. and Kumar H. (2018) Green Concrete: A Literature Review. International Journal of Engineering Research and Technology, 6(11), pp. 1-3.

[17] Huang X. and Kaewunruen S. (2020) Self-healing concrete. In: Samui P., Iyer N. R., Kim D. and Chaudhary S., eds. New Materials in Civil Engineering. Birmingham, United Kingdom: Butterworth-Heinemann, pp. 825-856. https://doi.org/10.1016/B978-0-12-818961-0.00027-2

[18] Institute, C. P. C. (2020), Curing of High Performance. Canadian Precast/Prestressed Concrete Institute- Technical Guide, pp. 1-20.

[19] Jackson K. (2018) Azure Magazine. [Online] Information on https://www.azuremagazine.com/article/sustainable-cladding-innovations/

[20] Jonkers M. H. (2007) Self Healing Concrete: A Biological Approach. In: S. v. d. Zwaag, ed. Springer Series in Materials Science: Self Healing Materials. Netherlands: Springer, Dordrecht, pp. 195-204. https://doi.org/10.1007/978-1-4020-6250-6_9

[21] K. K. et al. (2018) Self-Healing Concrete. International Research Journal of Engineering and Technology, 5(5), pp. 3817-3822.

[22] Kaja N. and Jaiswal A. (2021) Review of Precast Concrete Technology in India. International Journal of Engineering Research and Technology, 10(6), pp. 867-

872.

[23] Kamal Arif M. (2022). Precast Concrete Blocks in Building Construction: An Overview in the Book 'The Fundamentals of Building Materials', Edited by Gary L. Dixion, Nova Science Publishers, Inc., New York, U.S.A.

[24] Kamal Arif M. (2022) Stabilized Mud Blocks: A Sustainable Building Material, in the Book 'The Fundamentals of Building Materials', Edited by Gary L. Dixion, Nova Science Publishers, Inc., New York, U.S.A.

[25] Kamal Arif M., (2016) Recycling of Fly Ash as an Energy Efficient Building Material: A Sustainable Approach, Materials for Sustainable Built Environment, Key Engineering Materials, Trans Tech Publication, Switzerland, 692, pp. 54-65, https://doi.org/10.4028/www.scientific.net/KEM.692.54

[26] Husain A. and Kamal Arif M. (2015) Energy Efficient Sustainable Building Materials: An Overview, Sustainable Building Materials and Materials for Energy Efficiency, Key Engineering Materials, Trans Tech Publication, Switzerland. https://doi.org/10.4028/www.scientific.net/KEM.650.38

[27] Kamal Arif M., Brar T. S., Emerson P. (2014) Recycling of Construction and Demolition Waste Material for Energy savings in India, Materials and Technologies for Green Construction, Key Engineering Materials, Trans Tech Publication, Switzerland, 632, pp. 107-117. https://doi.org/10.4028/www.scientific.net/KEM.632.107

[28] Khazaleh M. and Gopalan B. (2019) Eco-friendly Green Concrete: A review. IAPE., United Kingdom.

[29] Kumar C., Rawat S. S. and Soni S. (2020) A Review on Self Healing,. IJARIIE, 6(1), pp. 447-453.

[30] Kumar E. P. (2020) n.d. civilrnd. [Online] Information on https://civilrnd.com/calculate-cement-sand-and-aggregate-for-nominal-mix-concrete/

[31] Kumar, G. M. and Moorthi, A., 2013. Properties of Green Concrete Mix by Concurrent use of Fly Ash and Quarry Dust. IOSR Journal of Engineering, 3(8), pp. 48-54. https://doi.org/10.9790/3021-03834854

[32] Labonnote , N ,Ronnquist ,A Manum ,B & Ruther P.(2016) additive construction state -of-the-art, challenges and opportunities. Automation in Construction, 72: 347-366. https://doi.org/10.1016/j.autcon.2016.08.026

[33] Lageman, T., 2019. Bloomberg. [Online] Information on

https://www.bloomberg.com/news/articles/2019-02-11/3d-printed-architecture-more-evolution-than-revolution [Accessed 14 March 2021]

[34] Le, T. T., Austin, S. A., Lim, S., Buswell, R. A., Gibb, A. G., & Thorpe, T. (2012)(a). Mix design and fresh properties for high-performance printing concrete. Materials and structures, 45(8): 1221-1232. https://doi.org/10.1617/s11527-012-9828-z

[35] Le, T. T., Austin, S. A., Lim, S., Buswell, R. A., Gibb, A. G., & Thorpe, T. (2012)(b).Hardened properties of high-performance printing concrete. In: B. E. M. Alexander, ed. Cement and Concrete Research. United Kingdom: Science Direct, pp. 558-566. https://doi.org/10.1016/j.cemconres.2011.12.003

[36] Lim, S., Le, T., Webster, J., Buswell, R., Austin, A., Gibb, A., & Thorpe, T. (2009). Fabricating construction components using layered manufacturing technology. Global Innovation in Construction Conference (pp. 512-520).

[37] Luhar, S. and Suthar, G., 2015. A review paper on self healing concrete. Journal of Civil Engineering Research, 5(3), pp. 53-58.

[38] Magaji, A., Yakubu,, M. and Wakawa, Y. . M., 2019. A Review Paper on Self Healing Concrete. The International Journal of Engineering and Science, 8(5), pp. 47-54.

[39] Mokal, A. B. et al., 2015. Green Building Materials - A Way towards Sustainable Construction. International Journal of Application or Innovation in Engineering and Management, 4(4), pp. 1-6.

[40] Munir, Q. and Kärki, T., 2021. Cost Analysis of Various Factors for Geopolymer 3D Printing of Construction Products in Factories and on Construction Sites. MDPI, 6(60), pp. 1-16. https://doi.org/10.3390/recycling6030060

[41] Nadarajah, N. (2018). Development of concrete 3D printing. Master thesis, Aalto University School of Engineering, Building Technology, Finland.

[42] Nematollahia, B., Xiab, M. and Sanjayanc, J., 2017. Current Progress of 3D Concrete Printing Technologies. Australia, pp. 260-267.

[43] Ozalp, F., Yilmaz H. D. & Ysar, S. (2018) 3D YaZici Teknolojisine Uygun Surdutulebilir Ve Yenilikci BEtonlarin [21] Gelistirilmesi . Hazir Beton Dergisi , Eylul 2018, 62-70

[44] Pacewicz, K., Sobotka, A. and Gołek, Ł., 2018. Characteristic of materials for the 3D printed building constructions by additive printing. MATEC Web of Conferences, pp. 1-9. https://doi.org/10.1051/matecconf/201822201013

[45] Papachristoforou, M., Mitsopoulos, V., & Stefanidou, M. (2018). Evaluation of workability parameters in 3D printing concrete. Procedia Structural Integrity, 10:155-162 https://doi.org/10.1016/j.prostr.2018.09.023

[46] Paul, S. C., Zijl, G. . P. v., Tan, M. J. and Gibson, I., 2018. A Review of 3D Concrete Printing Systems and Materials Properties:. Rapid Prototyping Journal. https://doi.org/10.1108/RPJ-09-2016-0154

[47] Rajput, K., 2021. Civil Jungle. [Online] Information on: https://civiljungle.com/m30-grade-of-concrete/ [Accessed 17 December 2021].

[48] Rehman, A. U. and Kim, J. H., 2021. 3D Concrete Printing: A Systematic Review of Rheology, Mix Designs, Mechanical, Microstructural, and Durability Characteristics. MDPI, 14(3800), pp. 1-43. https://doi.org/10.3390/ma14143800

[49] Renganathan, S., 2019. all3dp. [Online] Information on: https://all3dp.com/2/3d-printing-in-construction-what-are-3d-printed-houses-made-of/ [Accessed 18 April 2021].

[50] R, M. et al., 2019. A Research on 3d Printing Concrete. International Journal of Recent Technology and Engineering, 8(2S8), pp. 1691-1693. https://doi.org/10.35940/ijrte.B1134.0882S819

[51] Romdhane, L. and El-Sayegh, S. M., 2020. 3D Printing in Construction: Benefits and Challenges. International Journal of Structural and Civil Engineering Research, 9(4), pp. 314-317. https://doi.org/10.18178/ijscer.9.4.314-317

[52] Roussel, N., Rheological requirements for printable concretes, Cement Concrete Research, 2018, 112, 76-85. https://doi.org/10.1016/j.cemconres.2018.04.005

[53] R., Unknown. Conserve Energy Future. [Online] Information on https://www.conserve-energy-future.com/sustainable-construction-materials.php [Accessed 02 May 2021].

[54] S., 2021. civilplanets. [Online] Information on https://civilplanets.com/rate-analysis-of-concrete/ [Accessed 16 December 2021].

[55] Sanjayan, J. G., Nazari, . A. and Nematollahi, B., 2019. 3D Concrete Printing Technology. 1st ed. VIC, Australia: Butterworth Heinemann. https://doi.org/10.1016/B978-0-12-815481-6.00001-4

[56] Sir, C., 2021. Civil sir. [Online] Information on https://civilsir.com/how-much-cement-sand-aggregate-required-for-m25-concrete/ [Accessed 16 December 2021].

[57] Society, T. C., 2021. Concrete.org. [Online] Information on
https://www.concrete.org.uk/fingertips-nuggets.asp?cmd=displayandid=89
[Accessed 17 December 2021].

[58] Standard, I., 2007. IS456 - Plain and Reinforced Concrete - Code of Practice, New
delhi: Bureau of Indian Standards.

[59] Standard, I., 2009. IS10262 - Concrete Mix Design Proportioning -Guidelines, New
Delhi: Bureau of Indian Standards.

[60] Standard, I., 2011. IS 15916 - Building Design and Erection Using Prefabricated
Concrete - Code of Practice, New Delhi: Bureau Indian Standards.

[61] Suhendro, B., 2014. Toward Green Concrete for Better Sustainable Environment. In:
Procedia Engineering . Indonesia: Elsevier Ltd., pp. 305-320.
https://doi.org/10.1016/j.proeng.2014.12.190

[62] U., 2020. Sculpteo. [Online] Information on https://www.sculpteo.com/en/3d-
learning-hub/best-articles-about-3d-printing/4d-printing-technology/ [Accessed 14
March 2021]

[63] Unknown, 2021. Designing buildings Wiki. [Online] Information on:
https://www.designingbuildings.co.uk/wiki/Cast-in-place_concrete [Accessed 16
December 2021]

[64] Vijay, K., Murmu, M. and Deo, S. . V., 2017. Bacteria based self healing concrete -
A review. In: M. C. Forde, ed. Construction and Building Materials. Raipur:
Elsevier, pp. 1008-1014. https://doi.org/10.1016/j.conbuildmat.2017.07.040

[65] Wang, j., Wang, X. and Wu, P., 2016. A critical review of the use of 3-D printing in
the construction industry. Automation In Construction, 132, pp. 21-31.
https://doi.org/10.1016/j.autcon.2016.04.005

[66] Wangler, T., 2019. Digital Concrete: A Review. In: K. Baldie, ed. Cement and
Concrete Research. Switzerland, France, Netherlands: Elsevier Ltd., pp. 1-17.
https://doi.org/10.1016/j.cemconres.2019.105780

[67] Wolfs, R. J. M., Bos, F. P. and Salet, T. A. M., "Early age mechanical behavior of
3D printed concrete: Numerical modelling and experimental testing," Cement and
Concrete Research, Elsevier, 2018, pp. 103-116.
https://doi.org/10.1016/j.cemconres.2018.02.001

[68] Wu, J., Wei, H. and Peng , L., 2019. Research on the Evolution of Building
Technology Based on Regional Revitalization. MDPI, pp. 1-12.
https://doi.org/10.3390/buildings9070165

[69] Wu, P., Wang, J., & Wang, X. (2016). A critical review of the use of 3-D printing in the construction industry. Automation in Construction, 68: 21-31. https://doi.org/10.1016/j.autcon.2016.04.005

[70] Yang, H., Chung, J. . K. H., Chen, Y. and Li, Y., 2018. The cost calculation method of construction 3D printing aligned with internet of things. EURASIP Journal on Wireless Communications and Networking, Volume 147, pp. 1-9. https://doi.org/10.1186/s13638-018-1163-9

[71] Yu , C., Veer, F. and Copuroğlu, O., 2017. A critical review of 3D concrete printing as a low CO2 concrete approach. HERON, 62(3), pp. 167-194.

[72] Zitzman, L., 2018. bigrentz. [Online] Information on https://www.bigrentz.com/blog/the-future-of-building-materials [Accessed 14 March 2021]

[73] Zone, T., 2017. Youtube. [Online] Information on https://youtu.be/HfJpaRLJ0VI [Accessed 14 March 2021]

About the Authors

Dr. Tejwant Singh Brar is Senior Professor in School of Art & Architecture, Sushant University, Gurugram, Haryana, India and founding member of Architecture and Planning office Map Solutions, Patiala, Punjab India. He was awarded Ministry of Human Resource Development (MHRD) Institute Assistantship for pursuing Ph.D. in Architecture from Indian Institute of Technology (IIT), Roorkee, Roorkee, Uttarakhand, India. He is an Architect and Urban Planner having 23 years of Teaching, Research and Professional experience in the field of Architecture, Building Technology and Urban Planning. His areas of interest are Urban Water Resource Management, GIS, Remote Sensing, Urban Planning and Building Technology etc. He has published around 70 Research papers in various International Journals, Book Chapters and Conferences.

Dr. Mohammad Arif Kamal is an architect and academician having around 20 years of Teaching, Research and Professional experience in the field of Architecture and Building Construction Technology. He was awarded a Ministry of Human Resource Development (MHRD) Institute Assistantship for pursuing both M. Arch. and Ph.D. in Architecture from Indian Institute of Technology Roorkee (IITR), Roorkee, Uttarakhand, India. Dr. Kamal is presently working as an Associate Professor in Aligarh Muslim University, Aligarh, Uttar Pradesh, India. His area of research includes Environmental Design, Climate Responsive Architecture, Sustainable Architecture, and Building Technology, Traditional Architecture etc. He has published around 80 Research papers in various

International Journals and Conferences. He has published 3 books and 10 book chapters. Dr. Kamal is Editor-in-Chief of 5 International journals related to Architecture, and Building Technology. He has also edited 5 Special Topic Volume (Scopus indexed) related to Sustainable Building Materials, published by Trans Tech Publications, Switzerland.

Shubham Singh is an undergraduate student of Architecture at School of Art & Architecture, Sushant University, Gurugram, Haryana, India. He has done internships at Arcop Associates Pvt. Ltd. and Town and Country Planning Department, Gurugram. He is currently an Architecture Intern at Callison RTKL, Dubai. Shubham was also part of summer school held at Bartlett School of Architecture, London in 2019. His areas of interest are Architectural Design, Urban Design and Building Construction Technology etc.

www.ingramcontent.com/pod-product-compliance
Lightning Source LLC
Chambersburg PA
CBHW071500210326
41597CB00018B/2631